Cargo Theft Prevention

A Handbook for Logistics Security

Louis A. Tyska, CPP
Lawrence J. Fennelly

ISBN 1-887056-14-9

Printed in the United States of America

10 9 8 7 6 5 4 3 2

ABOUT THE AUTHORS

Louis A. Tyska, CPP, is a private consultant, expert witness, and chairman of the National Cargo Security Council. He is a member and past president of the American Society for Industrial Security (ASIS) and also served as chairman of its board in 1982. He has written, taught, and lectured widely in the security and business community. Together with Mr. Fennelly he has co-authored several books including *Security in the Year 2000 and Beyond, Office and Office Building Security,* and *150 Things You Should Know About Security,* and five others in that series. You may contact him at ltysk@aol.com.

Lawrence J. Fennelly, CPO, CSS, is president of Litigation Consultants, Inc. He is a security consultant as well as a forensic consultant. He is the editor and author of numerous books on security issues. Among these many books is the widely read *Handbook of Loss Prevention and Crime Prevention.* This work is currently in its third edition, and is highly regarded in both public and private enforcement. You are invited to visit his website at www.litigationconsultants.com.

CONTENTS

ACKNOWLEDGMENTS

Twenty years ago we wrote the grandfather for this millennium version of what we believe is a definitive work on the subject of logistics security. There are many who deserve recognition for their wisdom and efforts in the transportation field and their input to our efforts which appear in this work. Among those are: Joe Baker, Frederick C. Dorsey, Charles Girard, Jerald Jensen, John Kennedy, Art Kingsbury, Robert Macomber, John Mulattieri, Jack Publicover, Mike Shaw, Colin Venning, Brett Woods, Butterworth-Heinemann Publishing Co., and the U.S. Department of Transportation.

A very special thanks to the National Cargo Security Council, American Association of Ports Authorities, and U.S. Customs. Of course, none of our rewrite of this book would have been possible without Lucille Ham, our administrative support.

FOREWORD

For the past three decades, we have watched the estimated annual loss of transportation thefts rise like a balloon in the sky. It started in the early 1970s with a figure of over $2.8 billion annually, and has steadily risen to an estimated $15 billion per year. The tip of the iceberg, as many have described the figure, and so it is when cargo and transportation losses invariably result in claims which of their own, demand a specific course of action. The additional handling and processing of transportation claims will increase the amount of the losses further, and this is before a dollar estimate is placed on the inconvenience or replacement costs of the stolen merchandise.

What is an indictment against all of us who deal in the subject, is that there is no hard data available to rely upon. No one wants to admit to any degree of vulnerability or actual losses due to theft for obvious reasons. So, we continue to kite our losses in a series of ambiguous categories such as "mysterious disappearance," "lost in the system," "bill of lading/no freight," "lost on dock or in transit," or any number of such fuzzy references, all of which are everything but the proper reference.

The truth is that the time when our products are most valuable is when they are under the least amount of control. Security of finished goods or valuable componentry and supplies in transit is costly and time consuming. What is more disturbing is that due to pressure on management, generally, the security of merchandise in transit is not a very high priority.

The purpose of this book is to provide a broad based reference to transportation security. The terms logistics, transportation, and cargo are interchangeable in this book.

PREFACE

The Three C's of Cargo Theft

The established fact of cargo theft and pilferage is so commonplace that industry considers it part of the price of doing business. For many this is true, but there are those bright spots at the end of the industrial tunnel that have learned that good security controls enhance operational efficiency. Accept the challenge to reduce theft and pilferage, and in doing so you will contribute to overall profitability. If you are in the transportation, warehouse, and distribution business or interface with it, and we all do in some way, you know that there are risks involved. It is not acceptable to refer to losses as the, "price of doing business," or "mysterious or unexplained disappearances." To adapt such a fatal approach is to invite someone to rip you off. Such thefts are not accidents nor unavoidable, they are premeditated, well organized, and permitted. In the article, "Keep Cargo on Track," which appeared in *Security Management* in September 1986, we stated the following: "Lack of management control precipitates confusion, and confusion is the turned soil wherein conspiracies grow." That is more true today than it was yesterday as it pertains to cargo thefts.

The Three C's of Cargo Theft—when dealing with one or more cases of material and thus dealing with the more costly loss or claims as opposed to so called "petty pilferage" —are "Confusion, Conspiracy, and Common Denominator." These three ingredients have been observed in hundreds upon hundreds of physical and procedural security surveys conducted by the U.S. Customs Service Crime Prevention Program at seaports, airports, truck terminals, rail yards, and distribution centers. Observations of the three ingredients for freight and cargo theft were seen then as conditions conducive to such crime at the various facilities to varying degrees, but in *all* cases were always present.

Confusion

This ingredient, present at transportation, warehouse, and distribution points, is caused by the activities of the following:

- Personnel entering and exiting the specific facility: regular employees, casual workers, passengers, and crews, if applicable, repair and service people, and particularly truckmen.
- Movements of equipment such as trucks, cargo and freight handling equipment, containers, rail cars, aircraft, barges, and more.
- The varied and voluminous forms dealing with freight and cargo transactions such as passes, bills of lading, airway bills, manifests, equipment exchange forms, letters of credit, and a hundred more.

Generally speaking, for the sake of expediency, the movements of the above are uncontrolled and, thereby, contribute to internal confusion. Such a condition establishes an atmosphere of loose controls resulting in an impression of lack of management awareness which would provide the typical "action man" a favorable place for "doing business." In other words, a potential thief seeing poor controls weighs very quickly the odds of establishing a "contact" to assist him in committing a freight or cargo theft.

The judgment of whether or not management has tight controls over a facility can be felt immediately when approaching the point of entry or initial contact. Usually, this would be an entry gate or clerical office for freight or cargo. Lack of, or poor prescreening controls or procedures at entry, tip the odds in favor of the potential thief making the decision to steal.

Conspiracy

It has been defined in the legal field that a conspiracy is two or more people gathered together for the commission of a crime. My definition is slightly more specific where transportation and distribution crimes are established. A conspiracy can develop where truckmen, freight and cargo haulers, clerks and checkers, fork lift operators, supervisors, security personnel, or any combination of these people and others necessary to accomplish a successful freight or cargo theft in the confused climate of a transportation or distribution facility are present. Acceptance of the above eliminates the non-specific reference to loss when sustained as, "they hit us for a freight loss" or "they got us again." As Pogo once said, "We have met the enemy and they is us."

Common Denominator

The link which is a necessary piece of the conspiracy when the loss is more than one, bigger than a bread basket, and not as light as a feather is, of course, the dishonest truckman. We have stated in numerous lectures on the subject that you have a truckman involved when you cannot hide it under your jacket, place it in your pocket, stomach, or lunch pail. When dealing with other than pilferage from the warehouse, then it does occur that freight or cargo is hidden amongst legitimate freight or cargo movements or in the cabs of trucks. Hence, the truckman is to be controlled in his movements and activities while within a freight or cargo exchange area, in order to reduce risk and vulnerability.

Starting with the first ingredient of confusion, they can all be reduced and controlled with the initial declaration on the part of management that it intends to display a conscientious attitude and implement cost effective controls. This can be done with professional asset protection guidance to provide adequate physical entry/exit controls and internal security procedures. The source of the corrective measures starts with the same cause of the pervasive air of abandonment—MANAGEMENT. When management commits itself to seeing that confusion ends and control begins, then a reduction in costly losses and claims handling, poor customer relations, and related asset drain will be reduced and increased profitability will result. This need not be accomplished by large numbers of personnel or expensive hardware. The effective use of the existing organization and the correct assessment of the measures needed to improve conditions can be implemented, with the security professional standing beside the accountable manager providing him with the necessary expertise.

Part One

WAREHOUSE SECURITY

WAREHOUSE SECURITY

Security professionals generally agree that indications of theft almost always come about in one or more of the following three ways:

1. Inventory shortages (or variances).
2. Evidence or other symptoms discovered during facility inspections.
3. Information received.

These indications are the beginnings of a typical internal theft case, and I shall examine the first two items in detail in this chapter.

INVENTORY SHORTAGES

Many business executives hold to a time-honored belief that inventory shortages are valid barometers of just how a company is doing in relation to internal theft. Many seasoned security experts, however, are quick to point out that, although inventory shortages (occasionally referred to as "variances") may be a primary and basic indicator of internal security problems, such shortages (along with overages) are subject to a wide range of variables. Any of these can influence the picture to such an extent that "normal" conclusions may often be completely invalid.

It is not the intention of this text to cover in anything but a superficial way the accounting complexities involved in inventory problems. For those approaching the subject with no training in accounting, however, some primary explanations of this very important aspect of loss prevention are essential.

Book and Physical Inventories

Basically, there are two kinds of inventory: the *book* inventory and the *physical* inventory. The book inventory is arrived at as follows: At the beginning of each fiscal year, a company will have arrived at a figure of its inventory on hand by adjusting its book inventory to agree with the value of the physical inventory counted at year end. Throughout the year, certain purchases or production of additional inventory are made by the company. When received, these amounts are added to the beginning inventory figure. Also, sales of the company's products are made throughout the year, reducing the level of the company's inventory. Therefore, these sales must be subtracted from the other figures, which are normally augmented by purchases or production.

At the closing of the fiscal year, the company has thus arrived at a certain figure, which represents the beginning inventory, plus purchases and production, minus sales. This figure is known as the *book inventory.*

Once the book inventory is determined, virtually all companies are required to then make a physical count of all the inventory in their possession or under their control. This physical count may take several days and involve the efforts of many people. The figure resulting from the count gives the *physical inventory.* This physical inventory represents not what the company thinks it has on hand, but what it actually does have on hand. This count is compared to the book inventory.

Invariably, there will be unit differences between the physical inventory count and the book inventory figure. These differences can take the form of an overage or a shortage. Nominal differences or variances usually receive absolutely no attention, for they are considered normal for any business. Significant variances, however, whether they be overages or shortages, should be of concern to management.

Variances that are reflected as overages can be traced to a number of factors, such as price increases in commodities that went unrecorded in the books during the fiscal year, and also to poor physical counts.

Shortages can also result from a variety of causes, including internal theft, undiscovered external thefts, human error, and, once again, incorrect physical counts. Other variances can be caused by incorrect compilation of the book inventory amount and improper cut-off procedures.

Possible Weaknesses in Physical Inventory

It would seem clear, then, that any security executive attempting to study and analyze the results of a particular inventory should always be concerned with the adequacy of the actual mechanics of the physical count. There are a number of good accounting texts, such as *Retail Accounting and Financial Control*, 5th edition, by Robert M. Zimmerman et al., which cover the proper methods of taking a physical inventory.

The security executive should also concern himself with possible weaknesses in the system that have been caused by deliberate actions on the part of one or more members of the physical inventory team. One question that should be asked is whether or not the primary counters of the inventory are the persons who could have been responsible for theft of that particular portion of the inventory that they are now attempting to count. In numerous cases, key personnel have attempted to cover thefts throughout the year by padding inventory figures.

In most inventories, double counts are taken. It is essential that the security executive or accounting official make sure that there is no form of communication between the primary counting team and the second counting team. There have been occasions where the primary team, when finishing the count of a particular section of finished goods, would leave a number signifying the first count in a predetermined location in the section. This was done as a "favor" to the second team so that they would not have to make an actual physical count themselves.

As most inventory teams are constituted, each member of the team makes initial counts in certain sections and then serves as a secondary counter for other sections. If all the primary counters leave some kind of communication behind as to their count, then it becomes evident that no actual second count is ever taken. In this way, discrepancies between the first count and the second are never noted; thus a third and final count by members of management to verify a particular discrepancy is never made.

It is not difficult to visualize how an attitude of indifference on the part of certain workers can give rise to major inventory errors. One of the greatest deterrents to lax counting procedures by local personnel is a test count made by internal or external auditors. The mere presence of an auditor during the taking of a physical inventory certainly must have a beneficial effect. An auditor, who takes nothing for granted and who is aggressive and inquisitive by nature, adds significantly to the integrity of the whole process.

Inventory Undesirable in the Trucking Industry

In the trucking industry, very few companies use the term *inventory.* In the trucking business, to have any goods in inventory is to acknowledge that the freight is not being shipped. Only in a few instances would trucking terminals maintain inventories, and this generally would be items that are not immediately deliverable, such as items being held in a bonded section for customs clearance, damaged freight that has been refused, and other special categories. All-in-all, however, the thrust of the industry is to move the freight as rapidly as possible. Generally speaking, freight that arrives at the trucking terminal at the end of the business day is off-loaded from the city pick-up truck and then on-loaded to a trailer headed for its final destination. This trailer would then be dispatched, in all probability, by the following morning.

What inventory does exist, is handled by the OS&D (Over, Short, and Damaged) clerk. The OS&D clerk, depending on the individual company, may be required to make a survey of the terminal building on a daily basis; in all probability, he makes this survey at least weekly. The very nature of the job is such that this clerk quickly becomes involved in shortage claims by the customer, and plays an integral part in the attempted tracing of freight shortages.

INVENTORY IN MANUFACTURING

In manufacturing, there are two primary types of inventory with which the security executive must concern himself. These are (1) raw materials, and (2) goods in the production line, which are referred to as *in-process.* (Finished goods, which may be stored and warehoused at a factory location, will be covered under the heading Distribution: Warehousing and Trucking.)

Raw Materials and Componetry

In the case of raw materials, the basic accounting steps are similar to those outlined earlier. At the beginning of the fiscal year, the company starts out with a certain amount of raw materials. Additional raw materials received by the company throughout the year are added to this amount. Those amounts of raw materials that are withdrawn for purposes of production are then subtracted from these figures.

If there is good record-keeping and a sense of responsibility on the part of workers and supervision, the difference between book inventory and the physical inventory balances should not be large. At least it should be within acceptable tolerances, for there will always be a certain amount of waste in the production of goods from raw materials.

DISTRIBUTION: WAREHOUSING AND TRUCKING

In warehouse distribution, there are two major considerations as to inventory. First is the finished-good inventory, which results from the same basic accounting procedures discussed previously. Finished goods in a typical warehouse or distribution center should be influenced only by transactions that take place in the receiving and shipping departments. In other words, nothing should be added to the book inventory until it is actually physically received by the distribution center, and nothing should be subtracted from the book inventory until it is actually physically shipped out of the distribution center.

Concealed Shortages

The second area of concern to the security executive, and one that is not normally accounted for by standard inventory figures, is *concealed shortages*. A concealed shortage can be defined as a shortage of goods or inventory that is not obvious to the examining party.

Take as an example a liquor warehouse where cases of Scotch are being received. If a case has been tampered with and one bottle of Scotch removed, and the case shows no evidence of tampering, a concealed shortage exists. This may be discovered during the course of the fiscal year if the liquor warehouse has occasion to break open the case for the purpose of selling individual bottles to its customers. If this happens, the shortage is then noted and an appropriate adjustment can be made on the book inventory figures so as not to cause a physical shortage at the end of the year.

On the other hand, if the shortage is not discovered and the case is sold and shipped intact to a retail customer, the shortage is then discovered by the retailer. At this point, the retailer asks for a credit adjustment on his account with the wholesaler. The credit adjustment is usually made, but the adjustment does not have any effect on the wholesaler's physical inventory. Unless a separate record is kept of all credit adjustments made throughout the year for concealed shortages, the security executive never obtains a true picture of what is transpiring in a particular distribution center.

In the apparel industry, the same type of situation can exist where items are normally packaged three or four to a box. When the order-picker is filling a customer's order and withdraws one item from a particular box, it is general practice for him to adjust the number on the outside of the box so that a subsequent order-picker will know that the box does not contain the full allotment of garments. If this is not done, through negligence, then the customer is shorted and credit adjustment must be made. In the case of pilferage from these boxes, the thieves usually do not bother to make notations of the quantity reductions in the boxes, and thus concealed shortages are created.

Of course, security executives will immediately recognize that the problem of concealed shortages may very well be on the other end—in the retailer's receiving room. For this reason, spot-checks must frequently be made of boxes in a warehouse, as well as occasional double-verified shipments to customers. Generally speaking, however, the accounting and security departments must be guided by what they feel are "standards" for their industry in the matter of concealed shortages. Oftentimes, deviations from the standard can be the tip-off that something is seriously wrong, as in the following case.

> A moderate-sized shirt company was experiencing what appeared to be excessive concealed shortages in shipments of shirts to customers. As a standard, this particular company was compared to a larger shirt company within the same corporation that manufactured approximately three times the volume of the company in question. When these comparisons were made, it was found that the moderate-sized company had twice the amount of concealed shortages as its big brother, which was three times its size and in an identical business. With this kind of comparison, one can conclude only that there are major problems within the distribution or warehouse system itself.*

* Told to authors by a security director of a major manufacturer who wishes to remain anonymous.

A Security Benchmark for Common Carriers

Just as the retail industry is measured, from a security standpoint, by its shrinkage or inventory shortages as shown by a percent to gross sales, the trucking industry is measured by security for its amount of cargo claims.

Interviews with trucking executives have disclosed that the industry average is somewhere around 1.28 percent of cargo claims in relation to gross revenues. Some trucking companies have been known to come in at 0.39 percent, which is considered an industry low. To give the reader an idea of the tremendous amount of claims involved in the trucking industry alone, one company in particular processes between 18,000 and 20,000 claims per year, based on an average of 7,000 shipments per day.

RETAILING INVENTORIES

The very nature of the retail business dictates that inventory control and analysis are far more complex than for either manufacturing or distribution. In retailing, throughout the year there may be special sales and many other factors that, if not recorded and handled properly, will have an adverse effect on inventory figures.

Shrinkage and Reserves

In retailing, inventory variances or shortages are referred to as *shrinkage*. In virtually all retailing (and in some manufacturing companies) there is a psychological attitude that a certain amount of shrinkage is inevitable. As a result, the companies actually budget for a certain percentage of shrinkage throughout the year. In so doing, if their forecasts are correct, they do not close their books at the end of the year with a loss in inventory—a loss that would, of course, affect the bottom line of any profit-and-loss-statement.

These shrinkage reserves vary from store to store and from one segment of the retail industry to another, depending on the type of business. For instance, in the conventional department store, reserves may be set up that provide for inventory losses of from 1.5 to 3 percent of the net sales. In the discount or mass-merchandising segment of the retail industry, these reserves usually range from 2 to 4 percent. In clothing or specialty apparel stores, such reserves will usually range from 1 to 2 percent.

Shortage Reduction as a Contribution to Profits

Some security directors have targeted their programs toward these reserves in an effort to bring the inventory results in under the figure that has been reserved. If the actual shortage is lower than the reserved shortage, the difference can rightly be claimed as a direct contribution to the profits of the company. In other words, the company is bringing back into the profit column a certain portion of those reserves that were not utilized in covering for shortages that took place during the year. One major mail-order house, which also operates a large number of retail outlets, has always taken great pride in announcing this actual favorable difference as a contribution to the profit of the company.

Theoretically, if buyers, merchandise managers, and other executives did the proper, error-free paperwork; if sales clerks did not make mistakes with customers; and if all internal and external thefts were completely obliterated, then there would be absolutely no so-called "shrinkage." Accordingly, in the retail division of a company audited by Cluett, Peabody & Co., Inc., it has been publicly stated on many occasions that the goal of the specialty apparel stores in that division is a shrinkage of less than 1 percent. Several of the store groups within that division have achieved that goal for two or three consecutive years.

Shortages Based on Daily Count

In some segments of the retail industry, stores sometimes make a daily count of major clothing items. These counts are normally made at the end of the selling day and should correspond to the amount of sales of a major clothing item, such as women's dresses or men's suits, for that particular day. When problems are known to exist, and additional daily count is occasionally performed at the beginning of the selling days as well.

From time to time, these daily counts, although accurately reflecting the amount of shoplifting at a particular location, also may indicate problems within the store after closing. The following case history is an example of how a daily count inventory may add to the woes of the security manager.

> *Sherman Oaks, California.* Over a period of several months, the daily suit count of a clothing store showed significant discrepancies on an average of once or twice each week. In attempting to pinpoint

the discrepancies, the security manager instituted an additional count at the beginning of each business day. The additional count continued to confirm the previous pattern and indicated that suits were disappearing from the store during the closing hours.

The security manager undertook a number of lines of investigation, including a secret audit of the alarm company records, surveillance of the store, installation of a hidden closed-circuit television camera designed to operate during the nighttime hours, and other measures. None of these efforts, over a period of weeks, revealed anything, nor did a minute examination of the actual alarm system and physical construction of the building by an expert brought in for that sole purpose.

Next came a further review and audit of the daily stock counts. It suddenly was apparent that discrepancies were occurring in the morning counts that were being taken by a relief manager who worked in the store only one or two days each week. The morning count would be at variance with the preceding evening count taken by the regular store manager or his assistant. On those days, the evening count would always balance with that day's morning count.

Interrogation, coupled with polygraph examination, revealed a pattern of dishonest salesmen and the assistant manager padding certain customer purchases. Padding was accomplished by adding extra garments to legitimate items being purchased. The resultant shortages were then covered up by the store manager in the daily counts that he made, during which he "forced" the balances.

Although he was not personally involved in the thefts, the manager could not bring himself to admit, through his daily count, that there was anything wrong in his store. The discrepancies came to light only because the relief manager was making a completely honest count. In the meantime, thousands of dollars of men's suits had been lost through collusion. In addition, another considerable amount of money was expended for unproductive investigative efforts.*

INVENTORIES BASED ON ESTIMATED GROSS PROFIT

In many companies in the United States today, book inventories are not maintained in the manner described in this chapter. In the distribution field, because of the large amount of goods that are warehoused and distributed, upward and downward price changes are not recorded on the books throughout the year.

* Told to authors by a security director of a major manufacturer who wishes to remain anonymous.

One major distributor with nationwide operations, for example, stocked from twenty to twenty-five thousand different items of merchandise. It was virtually impossible to accurately record price fluctuations in book inventory throughout the year on every one of these products. Therefore, the company operated on what was termed an *estimated gross profit.* Estimated gross profits are simply arrived at by evaluating the actual gross profit percentages realized in prior years, and are based on what an intelligent appraisal of the company's business indicates will be the amount of gross profit.

Cost of Sales Category

In the manufacturing field, a similar estimate is sometimes referred to as a standard by which the company arrives at its gross margin of profit. In these situations, the basic accounting formulas described earlier are followed, except that a term known as *cost of sales* comes into play. Cost of sales in a manufacturing operation can be defined as all manufacturing costs incurred in finishing or producing a product. This would include raw materials, direct labor, and overhead costs. In wholesaling or distribution, cost of sales would usually include purchase price of the merchandise and any related transportation costs.

When there are no proper records being kept, price fluctuations or shortages in goods from the manufacturing unit to the warehouse are simply absorbed in this accounting category known as cost of sales. When this system is utilized, the only standard by which a company can judge its performance is an estimated gross margin or gross profit, based on past performances.

Built-In Theft Factor

The security executive must be acutely aware of the fact that, when estimated gross margins are used, there is always the possibility of a built-in historical theft factor. In other words, if there has been a level of theft that has remained fairly constant throughout the years, there would be no inventory shortages. It is only when people become greedy and begin to steal more than the historical amount that inventory shortages are discovered. The following case history is an example of an unusual situation where even an accelerated theft program did not show itself.

Seattle, Washington. A theft ring of six employees in a drug warehouse was suddenly discovered. The ring had been in operation for only six months and had been engaged in selling stolen pharmaceuticals to several drugstores in the metropolitan area. The ring had, by their own and two druggists' admissions, succeeded in smuggling out approximately $10,000 in pharmaceuticals in the six-month period.

The company closed its books for the fiscal year and at that time, took a physical inventory. Because of the fact that the company was operating on an estimated gross profit, not only did the $10,000 in pharmaceuticals not show up as an inventory shortage, but the company actually showed a small overage on its books!*

VARIANCES NOT CAUSED BY THEFT

As I have indicated indirectly throughout this chapter, there are many factors that can adversely affect inventory records, causing either overages or shortages. A comprehensive review of inventory problems must include some of the more common causes of variances that do not involve theft.

Price Changes

In the retail field, the changing of prices throughout the year is known as taking *mark-downs*. These must be recorded in the books. Failure to record mark-downs properly in the book inventory will invariably cause shortages at the end of the year. Other price-change considerations include mark-down cancellations, mark-ups, and mark-up cancellations. All these procedures, if not properly recorded in the inventory books, will affect the book inventory at the end of the year in an up or down manner.

Receiving Short Shipments

Another vulnerable area in any business operation is the receiving room. Security executives are well aware of the practice of a dishonest truck driver attempting to divert the attention of a receiving clerk so as to retain one or more cases on the truck while the receiving clerk is led to believe that the count is accurate and that he has received the

*Told to authors by a security director of a major manufacturer who wishes to remain anonymous.

correct number of cases in the shipment. This practice is sometimes referred to as receiving a *short shipment.*

Also, there are frequent occasions when a receiving clerk will discover that a shipment is short, but does not make note of it and fails to notify the accounting department accordingly. When this happens, the full shipment is put on the inventory books and an instant shortage is created.

A variation on this theme sometimes occurs whereby goods are received and then subsequently counted in a physical inventory, but are not recorded on the books until after the physical inventory is taken. This, of course, creates a "cushion" and can help to cover up other wrongdoing that has taken place during the year. By the same token, however, it also creates an instant shortage in the subsequent year.

Truck-Terminal Procedures

In the transportation industry, cargo theft is greatly minimized by a well-run trucking terminal. If control procedures are adequate, it is almost impossible for certain types of theft to take place.

Generally speaking, goods sent to a trucking terminal, either by the city pick-up truck or by a subterminal in the same region, are quickly off-loaded after their arrival. Generally, the driver of the incoming freight has no way of knowing who the off-loaders and checkers will be regarding the freight on board. In this way, unless the terminal is very poorly run and collusion can be set up beforehand, there is no possibility for a city driver to store contraband that he may have picked up during the tour on that particular day. It is for this reason that city drivers will often use the shipping and receiving docks of a particular customer account to store stolen merchandise picked up at another location. This is just one more reason why responsible members of management should always be alert to any freight or strange goods that appear to be concealed within their own facility on either a shipping dock or a receiving dock.

In the trucking industry, it is general practice for the outbound drivers not to make an actual physical count of the goods on their trailers. It is true that they have a manifest showing the destination point and goods to be delivered, but generally the trailer is sealed and the driver is not held accountable for the contents of the trailer.

Breakage and Damage

Another area not to be overlooked is breakage and damaging of goods by rank-and-file employees. Many employees, wanting to avoid reprimands by their supervisors, will fail to report broken and damaged items and will simply discard the evidence into the trash. When breakage and damage are not properly recorded so that book adjustments can be made, once again the inventory is shorted.

Unintentional Error

Last, but certainly not least, is human error. Indeed, this factor is probably one of the biggest of all. Unintentional underringing of sales to customers, the miscounting of goods in a customer's shipment, errors in typing, and endless other instances of human error may cause inventory shortages.

ADVANTAGES OF INFORMATION TECHNOLOGY PROCESSING SYSTEMS

In many companies today, the security executive can avail himself of a technology processing system to assist in determining trends in pilferage, etc. Many of these computer systems are used by companies for buying purposes only, and although they are not used for actual accounting records, in many cases they do constitute a perpetual inventory system. That is, on any given day, the computer may be able to indicate to a buyer or the security executive how much of a particular item of stock should be in the possession of the company and available for sale.

By simply monitoring the actual sales records of a particular item and then comparing them with the print-out of the data-processing system or perpetual inventory, the security manager is often able to verify suspicions and confirm that theft is currently taking place. It goes without saying that data-processing systems offer tremendous advantages in the control of internal theft as opposed to waiting for results of the annual year-end inventory.

In the transportation industry, especially with common carriers, computers are often used to maintain control of particular trailers enroute to a destination. Less frequently, they are used to maintain control over freight shipments and/or the generating of freight bills within the carrier's system. Additionally, computer data storage can also be used by the security executive to keep track of cargo claims and their relation to the gross revenues of the company

Facility Inspections

Most sophisticated security programs today include an ongoing plan of security inspection of the company's various facilities. Admittedly, most of these inspections are devoted to a review of procedures in order to determine compliance with company policy. The aggressive security manager can, however, utilize these periodic security inspections to look for indications of theft within the operation.

It would be impossible in the confines of this text to outline specific indicators for every type of business. Each type of industry is different, and only by learning in detail the nature of a company's operation can the security manager hope to be able to arrive at certain indicators of theft. For the purpose of illustration, however, I shall attempt to point out common basic points of inspection plus some specific examples of vulnerable areas in certain specialized industries. From these examples it should be possible to develop some insight into how particular guidelines or indicators of theft can be compiled for a particular company.

INSPECTIONS IN MANUFACTURING, DISTRIBUTION, AND FREIGHT TERMINALS—RECORDS OF OPENINGS AND CLOSINGS

Many facilities today are covered after hours by an alarm system that may be monitored from a central-station facility. When central stations are used, it is almost always possible to arrange to receive reports from the alarm company of any unusual openings and closings outside of normal hours. These are often supplemented reports prepared independently of the regular daily opening and closing reports provided by the alarm company.

In the case of installations with proprietary alarm systems, these systems may come equipped with an automatic printer or recorder that indicates not only the openings and closings of any doors in the installation, but also any tampering with the console or the printer.

For locations that cannot be covered by a central-station system or a proprietary system with a built-in recorder, a time-recorder lock can be installed on the main entrance of the facility. The time-recorder lock in turn is tied to sequence locks on the other doors. In this way, the security manager can obtain the same permanent record of unusual openings and closings offered by more sophisticated systems. This record can be invaluable when looking for a particular pattern of surreptitious entries or exits at the facility.

In addition to studying records of openings and closings for obvious holiday, Sunday, and nighttime openings, a security manager should also be alert to the extra two- or three-minute opening that occurs just after closing of the facility at the regularly scheduled time. Often, dishonest personnel have been known to place a cache of contraband near the exit and then reenter the building within several minutes of its closing on the pretext that they had forgotten a hat, umbrella, handbag, or some other item. This pretext is usually employed for the benefit of a fellow worker who may have left the facility at the same time as the key-holder.

When guards are on duty and shifts change at night, the situation becomes far more complicated. It is quite a common practice for some guards to arrive at their workplace anywhere from thirty to sixty minutes in advance of the relief time, and this may or may not precipitate an early departure by the guard being relieved. The problem, of course, comes about if there is collusion between the two guards and contraband is removed during the authorized opening of the door for the changing of the shifts. One solution for this possibility is periodic surveillance of the facility at these times.

Need for the Security Professional

In the trucking industry, many of the major carriers do not have any one person assigned to security responsibilities. What security functions are performed, are often performed by safety managers who make periodic visits to the various terminals and who include certain security checks along with the performance of their safety inspections. Although this is commendable as compared to no security effort at all, it is obvious that a security professional would have the unique security viewpoint on many matters and would be able to make a much more meaningful contribution to the welfare of the company.

Employee Locker Inspections

Employee lockers are another very productive area that should be inspected in detail. Although it is unusual to find any hard evidence of theft in such locations, the security manager can frequently pick up other clear-cut indications of violations of company rules and regulations. Often the major violators of certain rules can also be identified later among the dishonest employees of a location.

Alcohol. In particular, inspectors of lockers should be alert to alcoholic beverages, not necessarily in the original container. Many times, alcoholic beverages have been found in ordinary glass bottles, plastic containers, and insulated jugs.

Drugs. By the same token, many of the commonly used commercial drug preparations can often be found secreted in employee lockers. Here, though, the security representative must keep in mind that it is always possible the employee has a legitimate prescription for the drugs. This would be especially true for the so-called "uppers" (Dexedrine, etc.) or "downers" (barbiturates). On unidentified substances, the security manager should make an attempt to remove enough of the substance for laboratory analysis, but at the same time should attempt to photograph the remainder. He should also note the time, date, place, and any witnesses to the search. The most usual unidentified substance to be encountered will be the so-called "nickel" or "dime" bags of marijuana, or possibly cocaine.

Gambling Items. Gambling records or paraphernalia may also be found in lockers. There may be memorandums of IOUs or rosters of players within a gambling pool when identification numbers are being used. On occasion, the paraphernalia for various athletic or payroll check pools may be found.

Evidence of Theft. As stated earlier, it is the exception rather than the rule that hard evidence of theft will be found in lockers. But sometimes, packaging materials may be discovered that would indicate that a finished product had been appropriated for personal use. The most common disposition of such materials is usually into a trash container. On occasion, certain wrappings or other identifying labels are discovered in employee lockers, especially if there is some indication that trash containers are being checked by management on a daily basis.

Legal Considerations. Under the existing body of United States labor law, the rule has been laid down clearly in case after case that, when an employer puts his employees on notice that lockers are subject to inspection and will be inspected, it becomes a management prerogative to do so. There is absolutely no question of invasion of privacy or unlawful search involved at this point.

The one consideration that may play a part in determining the method used for locker inspections is whether or not certain agreements have been made with the plant union on this subject. Some agreements provide that the employee must be present when his locker is inspected, and if he is not, the fruits of any inspection cannot be held against the interests of the employee. In large plants, where it would be impractical for every employee to be present during a locker inspection, a compromise can often be worked out with the union to provide for a shop steward to act as an employee representative during an inspection.

At any rate, regardless of representation by or on behalf of the employees, locker inspections should be conducted without any prior warning and at irregular intervals. Private padlocks should be prohibited; lockers should be provided with locks that have duplicate keys for management, or even a master key.

Many security managers, not wanting employees or unions to become alert to the discovery of theft, will conduct their own private locker inspections after working hours, in a surreptitious manner. The object here, of course, would not be to make a case against a particular employee, but only to give the security manager some intelligence as to what may or may not be happening in the plant. Decisions on official inspections, which by their very nature precipitate action against offending employees, should be made in a careful and considered way.

Many trucking companies do not provide employee lockers, and when they do, there seems to be a great reluctance toward performing regular inspections of such lockers. This reluctance also is evident in most trucking companies regarding the inspection of employee automobiles. For this reason, employee automobiles are always restricted to an area outside the trucking terminal, within a separate compound. When this is not possible, most trucking firms restrict employee parking to an area away from the dock.

It is to be noted that the trucking industry is heavily organized by the International Brotherhood of Teamsters; this fact undoubtedly influences many of the operating decisions of management that may have an impact on security considerations.

Workstations

In most factories and distribution centers, employees have a particular area where they tend to accumulate minor personal effects over a period of time. Generally, this could be referred to as the employee's workstation. At many workstations, there are desk drawers to be examined, the rear of file cabinet drawers, small boxes containing personal items, or other containers. Here again, the security manager should be alert to any evidence of gambling, packaging material, contraband merchandise, or any form of illicit production records such as false shipping labels, false work orders, etc.

Sensitive Areas

Many plants and distribution centers have sensitive areas to which particular attention must be given during a security inspection. This would include special security rooms, security cages, and similar areas. Not only should desk drawers or file cabinets in areas of this type be closely examined, but also waste containers.

In the transportation industry, it is common for carriers to maintain a security cage within the terminal building. This security cage is ostensibly for high-risk freight items that cannot be delivered promptly, or those for which delivery has been refused. In some cases, trucking terminals have been known to maintain a particular trailer located at the dock and close to the office. They have designated this trailer as their security trailer for high-risk items such as described above. Often, this trailer, unlike the completely caged or screened-in areas mentioned above, is protected by nothing more than a heavy-duty padlock and kingpin lock.

Lavatories

Inspections of lavatories, both men's and women's, are not to be overlooked, for these areas often prove productive in the revelation of indicators of theft. Particular attention should be given to toilet water tanks, not only the inside of the tank but also the space between the rear of the tank and the wall; the area underneath wash basins; and any small access doors to plumbing traps, etc. Shelving, ventilation ductwork, dispensers of various types, as well as all types of refuse containers should be examined closely.

In a facility that stores or warehouses items of clothing such as hosiery, panty hose, and underwear, the inspection should be particularly alert to items of this type that may have been discarded in trash containers. It is quite common for women to discard worn panty hose in trash containers in the women's lavatory and simply replace them with a new pair from stock. Usually, along with the old discarded pair, the security manager can also find the original packaging material in which the new product was contained.

Full-Case Area

In distribution centers or warehouse areas, any security inspection should pay particular attention to the full-case area. If full cases are used for picking purposes in the filling of customers' orders, normally such cases will have been removed from the full-case area and can be found in an open-stock or order-filling area. The security manager, then, can conduct a meaningful inspection of the remaining full-case area.

Cases in the full-case area should be intact and sealed. Dishonest employees will sometimes deliberately damage a full case with the forklift truck or some other object merely to make it easy to retrieve contraband from the container. The security manager should be particularly alert to any apparent damage to the cases to ensure that the interior contents are still intact. In other areas of distribution, where customers' orders may be processed on the basis of three or four items to a small packing box, checks should be made of these containers to ensure that no concealed shortage exists.

Miscellaneous Areas

Other areas that should not be overlooked by the security manager during his periodic inspections are crawl spaces between buildings, air shafts, elevator pits, stairwells, roofs, and utility or steam tunnels.

In general, the security inspector should be watchful for any indications whatsoever within the facility of the existence of gambling, drinking on the job, or use of drugs. He should be alert to employees who appear to leave their workplaces for unauthorized reasons or proceed into unauthorized areas. Also, sleeping on the job as well as horseplay are often good barometers of the moral atmosphere within a plant. Many ingenious employees have devised methods of stacking full cases of merchandise so as to provide a hollow area within the stack in which they can drink, play cards, or sleep during the day. The security manager who knows his installation knows the practices of such people, and he can usually succeed in discovering their secret hiding places.

INSPECTIONS IN RETAILING

The retail store has its own special peculiarities to consider during a security inspection. As in manufacturing and distribution locations, these retail peculiarities can give clear-cut indications of possible theft or rule violations that give rise to eventual theft. The specific type of retail establishment will determine the particular indications of problems with which the security manager must be concerned. There are common denominators, however, and it is useful to examine some of the more generalized aspects of what to look for in such an inspection.

As in manufacturing and distribution, areas of concern in retail security inspections include merchandise hidden in unlikely places, poor housekeeping, and, of course, a pattern of unusual opening and closing of the facilities based on alarm company records.

Another area not to be overlooked, and one that would apply to any type of business, is any indication of tampering with the alarm system. Detection of tampering requires actual testing with the alarm company of every opening in the building.

Surprise Cash Counts

Most retail security experts agree that the first item of business during a security inspection should be a surprise cash count of the various funds maintained in retail stores. The security manager will be on the lookout for any shortages, in addition to postdated checks, IOUs, and other unusual items found occasionally in cash drawers.

The security review should also include a study of the house accounts of the employees to determine any significant delinquencies. An attempt should also be made to ascertain whether certain employees have had wage garnishments or have made repeated requests for salary advances. All these items can indicate that certain employees are in financial difficulties and should bear closer scrutiny on the part of the security manager.

"Hold" Merchandise

Virtually all retail companies have a stated policy on customer or employee "holds," or merchandise desired for purchase by a particular customer or employee at a later date. It is set aside, usually in a rear stock room, and held for a stated period of time. Upon the expiration of this period, the merchandise should be promptly returned to stock for sale to someone else. In order to control this type of situation adequately, most retail companies have specified that "holds" are to be clearly identified with the customer or employee's name and the date on which the "hold" was taken from stock and placed in a nonselling area. The security manager then should carefully examine all the "holds" bearing no names or dates, or those with grossly expired dates.

Cash-Register Irregularities

The retail security manager should be especially alert, during his security inspection, to "no sales" showing on a cash-register window—or, for that matter, any opening of cash drawers for no apparent purpose. This would also include the discovery of a cash drawer in an open position with no employee in immediate attendance. The security manager should also be alert to any blatant failure on the part of certain employees to follow house rules.

Proper Documentation

Two final areas that must be mentioned are a review of outgoing merchandise to determine whether any of it lacks proper documentation, and examination of merchandise in alteration departments to see that it has proper documentation.

Summary

In summary, the enterprising security manager in any type of business can utilize his security inspections not only to improve and maintain control procedures, but also to determine clear indicators of theft. In many cases, he may be able to identify specific employees who are involved in some form of dishonest program.

Loss Prevention Techniques for Cargo Theft:

1. Regular, unannounced inspections of employee lockers.
2. Regular, unannounced inspections of sensitive areas.
3. Prompt investigation of any freight shortage claims.
4. Personal involvement by top-line management in any suspected security matter.
5. Frequent and regular inspection of receiving and shipping docks by management of customer companies.
6. Covert surveillance of city delivery and pick-up trucks.
7. Overt surveillance of city pick-up and delivery trucks.
8. Guard patrol of terminal areas to prevent unauthorized intrusion and theft from dock areas.
9. Guard control of main gate area to ensure only the proper removal of loaded trailers, etc.
10. Periodic breakdown of loads to check on the presence of contraband merchandise.
11. Rotation of personnel within the truck terminal to ensure no pattern of certain persons loading or unloading.

WAREHOUSE SECURITY—25 RECOMMENDATIONS

Daily contact with the OS&D department (Over, Short, and Damaged) is essential. Its members are the frontline troops in the battle against losses. They should know of every shortage as it occurs. Each day, ask if any shortage has occurred of desirable or high-value products (power tools, clothing, cigarettes, liquor, etc.) and, if so, has the driver or warehouseman been questioned? The next day, inquire if the shortage from the day *before* has been cleared and, if necessary, question the employees involved. Security must be watching for date, time, commodity, name, and location patterns of current shortages and claims.

Claim-Theft-Loss Prevention Programs

Have an *effective* and *active* loss-prevention program. There appears to be a pattern to these programs. They start with a bang. Soon a paper jungle is created. Paper missiles are hurled back and forth among the prevention group, operations, and employees. What is really needed is someone free of paperwork. There is a place for films, talks, materials, and directives. But they do not replace continual day-to-day probing—checking on the spot, correcting on the spot, being on the warehouse platform one minute, and behind a truck in the city the next minute.

We must break out of the "wood and metal curtain"—the curtain of desks, files, and paper—once in a while. See what is going on around you. Companies today are split into "ghettos." Each frozen little island of employees sits, sometimes wallowing in a sea of inefficiency and apathy, moving paper rather than solving problems, living in their own "God's little acre." Employees in some departments not only never speak to those in other departments in their own company, but do not even know their names. Someone must be aware of what is going on in *every* department, and security, if it is to be effective, must know, must make liaisons, and must work with everyone. Its presence must be felt, not like a steel fist, but always there—gently, gradually, and strongly if needed, but ever present. Annual programs for security, safety, "be kind to a truck driver week" or any subject, are almost useless by themselves. It is the day-to-day continual battling with the problem, whatever it may be, that brings results.

1. All cargo claims should be on computer with a uniform code for cause, commodity, etc.
2. Issue a monthly analysis of cargo losses and base the loss-prevention program on the findings.
3. Support a strong, active, inquiring, searching, and relentless OS&D department. Security managers must cooperate with this group on a daily basis, looking out for shortages of high value and items desirable to thieves.

4. Top management must be involved in security and someone appointed by the president must have the overall responsibility for security.
5. Promote the idea of company spirit and employee motivation.
6. Insist on good supervision, especially at the foreman level.
7. Make each terminal manager responsible for the security of his terminal.
8. Have realistic company procedures and *enforce them.*
9. Work with the architect and have security in mind when planning new terminals.
10. Promote security awareness to *all* employees.
11. Records must be kept on all shortages, especially those with merchandise susceptible to theft. Watch for date, time, name, or any pattern that appears. These files should include information on the handling of all shipments, including:
    a. Who picked up the shipment (day and date picked up)?
    b. Who unloaded the shipment at the warehouse?
    c. Who reloaded it to the outbound trailer seal record?
    d. Seal record at destination.
    e. Who checked shipment at destination?
    f. What date and time was the shipment unloaded?
    g. Who was the delivery driver?
    h. What day and date was it delivered to consignee?
12. Maintain strict trailer seal control and accountability.
13. Ensure that the warehouse platform is under surveillance during lunch and break periods. This must also take place at shift changes.
14. Enforce strict control on interline drivers on the warehouse platform. Have their vehicles remain outside the premises until ready for unloading. When reloading, do not let them select their freight—provide supervision.
15. Keep strict control on the waybills and all documents that accompany freight.

16. Be on the lookout for billing errors and bring them to the attention of the billers.
17. Check your fences and lights regularly.
18. Keep a tight rein on all high-value full-load movements. Keep them moving. Secure with pin-lock when parked.
19. Establish good consignor-consignee liaison and, if they have security representatives, make personal contact.
20. Maintain strict control on all equipment (tarpaulins, power equipment, trailers, etc.) and account for all units regularly. These controls apply to both loaded and empty trailers.
21. Establish good hiring practices.
22. Keep security personnel in the terminal area on a roving check pattern.
23. Have a security cage on the warehouse platform for on hand freight susceptible to theft.
24. Security controls must apply to *all* employees, with management setting the proper example in honesty, truth, and justice.
25. Anticipate theft. Be on the alert when any particular article is in demand, high in price, and a target for theft. When there is a rail strike and extra freight comes your way, be careful. Create the "security presence."

WAREHOUSE DISTRIBUTION SECURITY

I have discussed previously, many areas of distribution as these areas relate to the warehouse. At this point, I will submit additional information on distribution security to be sure that I have touched most of the bases.

Trailers at the Warehouse

Each unit should have a serial number, trailer identification number on the sides, and large numbers on the roof for antihijacking purposes. Trailers should have their own parking areas, away from the employee's parking lot. Empty trailers should be separated from full trailers. Loaded trailers should be sealed immediately after being loaded at the loading platform. They should be checked for seal control, tire pressure, and temperature readings on refrigerated units. No trailer should be dispatched without proper documentation.

Trailers at In-House Units

The unloading process must be supervised closely. Drivers should remain with their rigs and not wander off throughout the store. Trailers should be unloaded as quickly as possible after arrival.

After a trailer is unloaded, the unit should be closed and resealed. In a split load, the new seal number should be noted correctly on the driver's delivery report. The store's receiving-bay door should be closed and locked, then resealed with a store seal. After a full load is delivered, the supervisor should check the trailer for merchandise under paper or pallets, or behind any interior equipment.

If the trailer was dropped overnight or if it cannot be unloaded until the next morning, the trailer should have a fifth-wheel lock placed on its kingpin to prevent removal. The trailer should have also a heavy-duty padlock.

High-value cargo packed overnight should be brought to the attention of the local police authorities with a request for periodic checks.

The receiving area should be adequately illuminated.

Trailer Security in Transit

A time study should be made of all regular runs. Regardless of traffic patterns at different hours on different days, a good study will predict time runs within a thirty-minute variation.

Security and safety teams should conduct periodic surveillances to determine traffic handling by the driver. The rig should be checked for unauthorized riders and unauthorized stops or routes. It should be checked for all safety features, for example, lights, horn, wipers, tires, brakes, etc. The team should be alert for any indications of unauthorized product removal or indications of seal tampering.

The driver should be checked for his Department of Transportation (DOT) driver log to be sure that it is not filled in before the run is made.

Background Investigations

The employment applications of all prospective drivers and warehousemen should be background checked. Their prior five-year work records should be verified with previous employers. One question you must ask a prior employer is: "Would you rehire?" Check the employee's driving record and be sure that his license is current.

Trailer Security at the Main Gate

Guards should check tractor cabs and trucks, (1) inside, (2) under the seat, and (3) under the sleeper.

Guards should check trailers for (1) doors that can be opened without breaking the seal because of poor sealing or because of removal of pins from the door hinges; (2) rear overlapping doors that can be pried open without disturbing the other door; (3) holes in sides or roof; (4) undercarriages; and (5) safety latches on handles that are defective.

Empty returning trailers should be checked for returned merchandise or pallets if you have a pallet-control project. Proper paperwork and seal-control procedures must be closely monitored.

Vendor Vehicles

Vendor vehicles should be issued gate passes in accordance with that procedure, and directed to the proper bay door.

Vendor vehicles should be checked on their way out for company product to the degree necessary. For example, for full-load deliveries there should be no product on board and the vehicle should exit empty. On partial-load deliveries, the unit should be examined for foreign product—that is, product belonging to the warehouse but not to the vendor. The partial load remaining should be as described by the receiver and noted on the gate pass. The gate pass should be stamped with date and time and filed for six months.

Vendor drivers should not be permitted away from their vehicles during standby or unloading time. They should be held responsible for their helpers or any casuals hired to unload. After unloading, they should be encouraged to leave the property immediately.

Distribution Security Aids

A fifth-wheel lock is a heavy-duty lock collar that is placed around the kingpin of a trailer (called the fifth wheel) when the trailer is detached and dropped from a tractor. It is key operated with an interchangeable cylinder core made of an alloy with a high melting point. It is used in high-crime areas where there has been a history of hijackings. It prevents another tractor from hooking up the trailer.

A tachograph is a device that is attached to the wheels and odometer. It produces chart graphs of the vehicle's activity. Under proper conditions, you can tell a driver's mileage and stops after reading the charts.

Electronic alarm sensors can be installed that go into alarm if the vehicle is moved or broken into. The variety available is comprehensive, but suffice it to mention that the devices can be installed on the tractor, truck, trailer, or nearby to detect unauthorized movement.

There are many types of padlocks available. Be sure that a padlock is case-hardened steel and that the key is difficult to duplicate. Use interchangeable cores so that cores can be changed periodically without effort or added expense.

Roof-top markings—trailer identification placed on top of the roofs for easy helicopter identification—are immensely helpful in hijack situations.

SATELLITE TRACKING

Intermodel Transportation and Shipment Control

We didn't know it at the time, but one of the benefits of the space race, the cold war, and President Ronald Reagan's Star Wars, were the advanced systems of communications and other specialty function satellites in earth orbit. These satellites are now providing benefits outside the military arena. They are used for weather forecasts, broadcast TV, telephone service, maritime navigation, search and rescue, and a whole host of civilian uses. A widely used system for navigation is the Global Positioning System (GPS), which uses a constellation of satellites operated by the U.S. Department of Defense. There are also an increasing number of privately funded and launched satellites in orbit. One of the ancillary uses for these systems and their satellites is the tracking and monitoring of high value assets, when they are at risk within the various forms of intermodel transportation in use in global commerce.

How Does GPS Work?

The fundamental basis of the GPS is a triangulation process, where a GPS receiver measures its distance from three satellites using the travel time of radio signals broadcast from the satellites to provide precise location information. The GPS receiver uses the signal from a fourth satellite to remove any errors in its local clock, allowing for very accurate timing. This timing process allows for the highly precise measurement of distances that gives GPS its accuracy. Along with distance, the exact location of the satellite in space is also needed. Constant monitoring by ground based radars allow for very precise position updates to be sent to the satellite for this purpose. The satellites broadcast a signal, which contains information on its location, velocity, and the time of transmission based on very accurate atomic clocks. A GPS receiver that can access signals from three satellites for triangulation and a fourth to correct the receiver's clock, can locate its position on the globe with excellent accuracy, from 15 to 30 meters. In fact, if you use a differential GPS, one that removes the built-in error placed in the system by the military to prevent malevolent use, by using a additional fixed land based signal, you can achieve accuracy in the 3 meter range.

Other Systems

Some satellite tracking and monitoring systems do not rely on GPS. One type of system which has been in use for a long time, uses commercial satellites to provide location and communication information for high value assets. This information is sent to a proprietary network management center for distribution to customers. These systems are primarily used for large fleet tractor trailer logistics management with security applications as bi-product. These systems can report on driver performance, route integrity, unauthorized entry, seal tampering, and just about anything else that unrestricted imagination can conceive.

Another type of system that is used to provide access to containers, trucks, and ships anywhere in the world in near real time, uses the Low Earth Orbit (LEO) ARGOS satellite system. Data distribution is provided via low earth orbiting satellites which circle the earth several times a day. Sensor data and GPS positions are uplinked/transmitted and subsequently downlinked/transmitted to the end-users' operations center.

Recent Developments

Currently, the GPS is experiencing a boom in commercial systems that use and benefit from precise location data supplied by this methodology. One would think a technology that can track elephants in Malaysia and sea turtle migration in the Gulf of Mexico would be applicable for use in security applications, such as remote vehicle tracking and protecting high value containers in the intermodel system. As with any new application of technology, these protective systems are not for everyone, but must be carefully analyzed for cost benefit before application. Costs can be prohibitive, but advances in Global Positioning System technology and digital communication systems are rapidly converging to provide affordable tracking for high value cargo and other mobile corporate assets, including personnel. The new systems in some professional's view are improvements over systems using up and down link transmission to earth orbiting satellites. They combine GPS receivers and cellular telephone transmitters in a single unit, sometimes at the circuit board level. The units utilize GPS to identify the current location of the receiver and then transmit the coordinates to a central station via the cellular telephone system. Central station software will give the position in latitude and longitude and depending on the type of graphic display and map available, display the location of the receiver. In addition to location, information from other sensors monitoring cargo status can also be transmitted to the central monitoring point directly. It must be noted that GPS cannot always receive satellite signals to achieve position information. Urban settings with tall buildings and garages can degrade or block signals. Additionally, cellular networks also may have areas of poor coverage.

All these systems described have advantages and disadvantages, when used in the various locations and conditions which can be found in intermodel logistics.

Trucking—Ground Transportation. One supplier of low earth orbiting satellites permits its clients to track, monitor, control, and communicate with their equipment anywhere in the world. This can be accomplished through a variety of ways. The flexibility of being able to choose between leased line, dial up, or Internet access to the information allows the end user to access a dedicated line to the service satellite. The system we refer to here is run on a Windows-based Pentium PC wherever the client elects to establish a control point. Simply stated, the rolling stock, wherever it's deployed, transmits its own signal to the providers' orbiting satellite. The equipment is sending from its own battery operated transmitter via its gateway with the service provider through a private network to the client's gateway located at the monitoring site via selected or available means. The battery has a two-month life between charges and can recharge from the power unit/Tractor when loaded. The 12-volt DC battery has a 60-month life. There is a requirement of a four-foot horizontally mounted and grounded antenna. The transmitter placed within or on the asset being tracked must be a self-contained low power one, which accepts both digital and analog sensor inputs. A good one will have a microprocessor, which is powerful enough to do both data logging and or transmit. All of the above applies to tractor and trailer trucking. There are slightly different circumstances for refrigerated units, the most significant of which is that the transmitter is generator powered for obvious reasons. Another difference is the antenna that is vertically mounted on the unit, which can result in some operational limitations.

What all of the above permits, is the client to track ground equipment in near real time indicating progress in its routing, traveling within its route, and estimated time of arrival. Consider the benefits of learning the unit has strayed from its assigned route or its temperature has changed in a controlled climate unit. When selecting a supplier with whom to purchase or lease equipment, be selective as the "real-time" notifications can be up to six hours in some instances. Hardly real time.

When referring to ground transportation it must be accepted that, of course, rail transportation is included. All that would apply to trucking and containers is apropos to rail. However, there can be slightly different types of threats. The conditions under which rail movement are subjected to in their routing makes satellite tracking ideal. An example of the effective alarm signal received via satellite and then down-linked to a particular rail operator's monitoring station can be found in the following incident: It happened one night that a rail movement had to spend some time on a sidetrack adjacent to a residential neighborhood. As the tranquillity of the night became disturbed, along with the sleep of the residents, the unyielding drone of a string of refrigerated railcars caused a mob reaction. The residents, in a variety of attire, assailed the unsuspecting cars choking off the generators to halt the noise and, thus, allow them to return to sleep. The satellite tracking system would have none of that. It dutifully received the signal that there was trouble with the refrigerated units and notified the client.

Ocean Transportation. Global transportation conditions are complex and no method of transportation should be excluded. In some parts of the world, they are still bringing material to be exported by vessel to the docks via donkeys. No, we aren't suggesting that we strap a transmitter receiver on a donkey for GPS tracking. However, there is a frequently used method for water borne transportation that's overlooked but often used in many parts of the world—the barge.

In addressing marine tracking, it's pointed out that the satellite systems described have worked effectively with barge tracking as well as with ground transportation. There are some problems when we enter into the movement of containerized goods in transit. Basically, the methods are similar as with ground transportation. The concerns are that not all the containers are deck loaded. Somebody's containers have to go in the lower hold of a container vessel. Consider the transmitter problems associated with the storage location or loading position of containers. Requesting specific loading priorities in water borne transportation are generally not an option, nor dependable if agreement is negotiated. Developers and research efforts are attempting to solve this problem through the use of an on-board network, which would receive the individual unit data on command and uplink the information to the orbiting satellite where it would be analyzed and distributed to the appropriate client. Also to be considered is the "portability" of the equipment used to establish and interact with satellites. Not all containers are committed to exclusive use. Can you track any container? Can you monitor for intrusion? Can you remotely trouble shoot when looking for problems in refrigeration movements? The answer to all is yes! However, is it consistent? No!

Air Transportation. We think it best to describe an interview with a West Coast based security director who exclusively uses international flights to transport high value product. We wanted the details of the GPS system, which integrates positioning technology with cellar technology. We were curious because of our experiences on aircraft when the flight attendant comes on the PA system to say, "The use of cell phones, personal computers, and other similar electronic equipment is prohibited during take offs and landings!" We had no idea at the time, but our thinking was instep with this particular security director's CEO. The security director described in detail working along with the knowledgeable engineer and developer of equipment that uses cellular technology integrated with GPS tracking, to develop a system for monitoring the corporation's small plastic containers. The general value of the contents per container generally falls between $80,000 and $90,000. Frequently, their shipments consisted of more than one container per shipment. All shipments were via international flights, with the calls from the containers programmed to advise the monitoring station at the security director's office of the exact location and status of the shipment. No problems. The system was tested and about to be implemented when the CEO asked what if the system calls during take off or landing? The fear of any association with the remote possibility that this could happen, stopped the program from implementation. Faced with this concern the supplier engineered a solution. Photoelectric cells were placed within the system and used to shut down the calls from the containers when loaded on board the aircraft in the cargo hold. The program was resubmitted with the correction and re-explained. It was still rejected since cargo holds do have lights and, at times, they are left on. Anybody want a slightly used tracking system for $10,000?

Investment

It must be understood that with any emerging technology when first applied in the field, there will be almost as many different costs associated with its implementation as there are suppliers or vendors. Consider if you will, how the end users decision relative to type of monitoring they elect to have, could impact on the costs. If a client elects to have status and/or location data hourly as opposed to being advised for specific exceptions when they occur, the costs will be considerably different. Currently, some suppliers are making their unit transmitters installed available between $300 and $400 with exception and location tracking approximately $20 per month. Others charge by the downlink notification at 60¢ per call. That's considerably more than $20 per month if you elect hourly notifications. It could end up costing over $5,200 per year, per unit. It would be cost prohibitive under such expense. What is required is a risk and vulnerability assessment to determine a reasonable and realistic notification and monitoring schedule. Where have you heard that kind of thinking before?

Is satellite tracking the future for transportation and distribution logistics? You bet it is! It does have its problems. However, as technology development and operational transportation and logistics functions interact and adjust, learning from each other, we will see seamless implementation and results. The one factor we don't know is how the global cargo thieves will react. The game of chess goes on.

GPS Web Links

http://www.inmet.com/~pwt/gps_gen.htm
http://www.gpsworld.com/resources/links2.htm
http://www.irscorp.com
http://www.trimbal.com
http://www.qualcomm.com
http://www.seimanc.com
http://www.orbcomm.com

Location Location Location

Wireless Enhanced 911. Since the first cellular telephone system was activated in 1983, there has been a tremendous growth in wireless services. Industry professionals state that approximately 20 percent of the U.S. population or about 45 million people have cellular telephones. This growth has, in general, added to public safety as the cellular telephone is used more and more as a tool in emergency management by public safety agencies and by business organizations. With the robust expansion of wireless services, cellular telephone calls to emergency services are increasing. In fact, wireless calls to 911 represent about 25 percent of all 911 calls placed annually or about 50,000 calls per day. It is estimated that by the year 2002 the number of calls daily may reach 130,000 almost equaling the number of 911 calls received by wireline/landline.

In the past, processing these calls has been a difficult challenge to both the wireless carrier and the public safety agency. Wireless carriers faced the problem that most cellular switches were not able to pass the callers phone number to the 911 telephone switch to perform intelligent routing, a fact that is further complicated by public safety agencies being organized along state, county, and municipality lines. Cellular switches do not recognize these divisions and often would pass the calls to a single public safety agency. Even when luckily handed off to the correct agency, unlike landline calls, these calls did not provide call back numbers or the caller's location. Often, when callers remained on the line, because many were in "transit" in areas unfamiliar to them, they were not able to provide precise location information. This would require the dispatchers to spend extra time with each call simply to determine location. In general, location information on a landline call in enhanced 911 situations is displayed to dispatchers in just a few seconds. A wireless phone call often would take several minutes for dispatchers to obtain location information, if at all.

Why Is This An Issue

Some high profile incidents come to mind. In a January blizzard in South Dakota, Karen Nelson was trapped in her car by heavy blowing, drifting snow in sub zero temperatures. She was saved by calling for assistance on her cell phone, but it took rescue teams over 40 hours to locate her. A few years ago, rescue efforts were delayed when responding to a train wreck near the meadow lands of New Jersey when the first call for assistance was received via the wireless network. In fact, a simple key word search for news stories containing the words cellular telephone and rescue will reveal dozens of articles from papers all over the country. The wireless phone certainly performs a life saving function, but locating the person requesting assistance is sometimes a major effort.

Rules

As a result of the issues noted previously, a proposed rule by the Federal Communications Commission (FCC) requiring *all* wireless carriers to provide location information to E911 public safety services whenever a wireless user dials 911 was considered. In June of 1996, the FCC issued a Report and Order in Docket 94-102, formalizing the requirements and establishing an implementation schedule for wireless E911 systems.

Phase One was completed April 30, 1998. It required wireless carriers to report call back numbers (Automatic Number Identification or ANI) and the cell site or sector.

Phase Two to be completed by October 1, 2001, requires wireless carriers to report the location of calls (Automatic Location Identification or ALI) with an accuracy of 125 meters for 65 percent of callers requesting to 911.

Solutions

One of the technologies that is being used to solve the location problem is known as Time Difference of Arrival (TDOA). This technology uses precise measurement of the time when a cellular phone transmissions reach different wireless signal collection sites. These differences are used to geometrically calculate the location of the transmitter. Now this is certainly not the only technology that can be used, others are Power Level Attenuation and Measured Angle of Arrival (direction finding) but it is one that has some advantages for ease of implementation and has undergone testing by the State of New Jersey. Results of the testing as published in *Report On The New Jersey Wireless Enhanced 9-1-1 System Trial State Of New Jersey Department of Law and Public Safety* were very encouraging, with participants in the trial holding a consensus opinion that the wireless location system significantly improved their ability respond to emergencies in a timely and efficient manner. Other applications mentioned for this technology are location sensitive billing, vehicle and fleet management, and inventory and package monitoring.

All technologies have limitations and those who feel this will lead to small, portable personal panic alarms will, for the near future, perhaps only be half right. Wireless networks do have locations of poor reception and 125 meter accuracy in a dense urban locations may not completely solve the location problem. In fact, TDOA relies primarily on the first burst of information (data) from the phone to compute location, those who remain in motion will cause location errors greater than 125 meters.

However, most certainly, when location reporting technology is widely implemented by wireless carriers, emergency response services provided by public safety agencies using enhanced 911 service will be significantly improved.

Part Two

AIRLINE SECURITY

AIRLINE SECURITY

THE AIRLINE INDUSTRY RESPONSE TO CARGO THEFT

Because this book will be read by security professionals in other industries, this section can serve a dual purpose. Security people in other industries may use some of the principles and concepts that are covered here. Others may suggest additional or better means for the airline industry's approach to defeating cargo thieves. None of us is as smart as all of us!

THE AIR CARGO SECURITY PICTURE

A Unique Environment

A large airport is really a small city with its restaurants, bars, barber shops, other concessions, buses, taxis, hotels, banks, and police and fire departments. From a security viewpoint, the airport has many advantages that are absent from the environment of a city.

1. It is divided into a public area and a nonpublic area. The nonpublic area has a fence around it.
2. Entrance to the nonpublic area is controllable.
3. Personnel and vehicular identification procedures may be used.
4. There are more law-enforcement officers on duty at the airport: the regular police complement, those supporting the antihijacking program, U.S. Customs officers inspecting baggage and cargo, postal inspectors concerned with the security of the mails, Immigration and Naturalization Service inspectors, and, at some airports, FBI resident agents.
5. Airports are not usually located in a high-crime or high risk areas.

6. The airlines are the principal tenants at the airport and have similar problems.

The disadvantages of an airport environment are fewer.

1. Many airports have an addition here, a new building there, with little or no regard for designing security features into construction and modification.
2. Air cargo is carried on freighter aircraft or in the belly of combination passenger-cargo aircraft. There are no special problems in loading or unloading freighter aircraft. There are unique problems, however, in loading and unloading the cargo on a combination aircraft. Scheduled aircraft carry passengers, hand baggage, checked baggage, commissary supplies, mail, and freight. The passengers and their hand baggage travel the same route to the aircraft, the other items travel different routes to arrive at the same destination. Thus, there are five avenues to the aircraft that have to be secured, and there must not be a delay.

Other infrequent problems may occur. Cargo may be offloaded at the wrong stop or not offloaded at the correct stop. A plane may encounter mechanical difficulty requiring offloading and return to the terminal. It is axiomatic that cargo not in normal channels is the cargo that is more susceptible to theft or pilferage. A former security representative of Swissair puts it in a more positive tone. He says that "the safety factor of a consignment goes up as the time required to reach destination goes down."

Type of Cargo

Air cargo differs from cargo carried in other transport modes in that much of what is carried by air meets the requirements of a fence—it is portable, valuable, and marketable. On average, the value of an airfreight shipment is 50 times that of a rail consignment and 30 times that of a shipment by truck. Accordingly, it is essential that sound security procedures be followed. The loss of a single shipment can cause the claims-paid ratio to skyrocket.

The Cargo Revolution

The conversion of the passenger jet to carry cargo began in 1979. These air freighters include the wide-body jets such as the Boeing

747, the Douglas DC-10, and the Lockheed L-1011. Each time an airline buys one of these larger advance-technology passenger planes, it is also buying about two-thirds of a conventional jet freighter—on the basis of belly cargo capacity. As a result, many shippers began to ship more and different commodities by air, with the result that carriers had to begin constructing new cargo terminals to handle the increase.

With the jet came all sizes of containers for transporting air cargo, and, from a security person's viewpoint, the containers arrived none too soon. Containerization is an effective security measure. Individually packaged, uncontainerized goods in transit, particularly if they carry the product's identifying label, are tempting targets for thieves. Containers conceal the nature of the goods. Hence, containers remove much of the theft temptation. If the container is locked and sealed with a registered and numbered seal, and the seal number is on the airbill, the security of the container's contents is further protected.

Another headache is the ransacking of registered mail by thieves working out of the principal airports of the United States. There have been many estimates of the value of the contents of the stolen pouches. Probably the most factual information came from an official of the Postal Inspection Service in a conference of airport, airline, and federal officials who declared: "From 1967 to 1970, over 500 pouches containing registered mail sent by air disappeared at airports in this country—mail which contained cash, jewelry, bonds, and securities worth over $80,000,000." This quote is from a speech by Carl E. Lawrence, Assistant Chief Inspector, Security, U.S. Postal Service, at the Annual Workshop for Chairmen of Local Air Transport Association Security Committees, Houston, Texas, April 28–29, 1976. This was the last study of its kind in such depth known to the authors.

Policy Agreements

The air-carrier security directors agreed on four basic security concepts.

1. *The name of the game is prevention.* Despite the fact that the ranks of security directors include many top-flight investigators and former law-enforcement officers, there was unanimity on this concept. They realized that investigation, apprehension, prosecution, and sentencing have a role to play in crime deterrence, but the primary thrust must be on preventing the crime

from happening—the age-old precept, take away the attitude and opportunity to commit a crime.

2. *Security is noncompetitive.* This is a highly unusual concept for a highly competitive industry. However, all agreed that making competitive claims about security is not the best approach in marketing air transportation. In addition, carriers agreed that if one carrier developed some new procedure, control, or device to enhance security, it would be willing to share this improvement with their colleagues in other airlines.

3. *A security awareness must be developed in airline and airport employees.* There was a consensus that this might be the most difficult of the concepts. The airline industry was built on flight safety, sales, service, convenience, courtesy, the right word, and the nice smile. To the majority of airline employees, security consisted of safes, locks, and contract guards. This type of thinking had to be changed.

4. *It is essential to have security with facilitation.* The U.S. scheduled airlines transport millions of passengers annually, and millions of tons of cargo. Hence, there was complete agreement that an air carrier cannot have Fort Knox security and run an airline. Each flight segment is part of a carrier's route system. Security-caused delays at point "A" could cause a serious disruption in the system.

Colin Venning, when he was Director of the Security Council of the Ontario Trucking Association, expressed the feelings of many security officers when he wrote, "The battle cry is PREVENTION—Enforced controls is the sword—awareness is the shield. Security must be in the air—not violently—not suddenly—just there—ALWAYS THERE."

Vulnerability Assessment

The security directors analyzed the security vulnerabilities and found that they could be placed into four categories: administrative, personnel, physical, and operational.

Administration. Administratively, two large voids were discovered: the lack of a central point where cargo-security activities at the nation's airports could be coordinated, and the lack of an organization that could explore methods to improve all aspects of cargo security—a laboratory for innovative security measures where proved security procedures could be combined with new ideas. To fill the first void, the carriers established the position of Director of Security of the Air Transport Association (ATA). To fill the second, foreign and domestic carriers serving the metropolitan New York airports of Newark, La Guardia, and Kennedy joined in 1968 to form the Airport Security Council. Liaison between the Director of Security of the ATA and the Airport Security Council is very close.

Personnel. As an early step in the 1970s, the airlines analyzed the vulnerability of personnel security. The analysis revealed that some carriers did not have good personnel security programs. Steps were quickly taken to ensure security input for the personnel hiring process. Also introduced were security briefings for new employees and a security education program for all employees, with stress on the security responsibilities of supervisors. The concept of the exit interview of those who terminate employment has been used, again with heavy emphasis on eliciting reactions to the company's security climate.

The most difficult, and probably the most important, personnel security need was to develop security consciousness in employees whose training was limited to customer service. Steps taken to develop security awareness will be included in the section of this chapter dealing with the work of local ATA security committees.

Physical Security. Most airports were not built with security in mind, and many airports have grown with an addition here, a new building there. Security was left out of the design process. To fill this void, a task force of the ATA Security Committee prepared and distributed a paper on airport security design criteria. These criteria set forth security design standards for guidance in the building of new airports and modification of existing airports.

The security of airport parking lots presented a special problem because, at some airports, the service roads to and from cargo areas run directly past employee parking lots. Such a situation runs counter to taking away the attitude and opportunity to commit a crime.

A third physical security consideration was the security of the cargo facility itself. Once again, a task force of the ATA Security Committee prepared an airline security checklist covering location, lighting, fencing, police patrols, control of truckers and visitors, door control, changes in the facility caused by weather conditions, key control, and other aspects of physical security.

Air-Freight Operational Procedures. Here was where the security directors made optimum use of the Airport Security Council. All questions on the best way to run a secure terminal were referred to the Council. Procedural guidelines were prepared by the Council ranging from the role of packing and handling in cargo security to shipper cooperation.

LOCAL AIR TRANSPORT ASSOCIATION (ATA) SECURITY COMMITTEES

Developing Security and Awareness

The ATA Security Committee found that nearly all the security problems were at the large hub airports. What was needed most at these airports was a vehicle that could develop security awareness in all employees at an airport. A secondary need was a vehicle that could bring together all interested parties to analyze crime patterns at the airport and to coordinate preventive measures. The carrier security directors decided to establish local ATA security committees at twelve of the largest airports. The immediate success of the original twelve caused expansion to thirty-three and then to fifty airports.

Each committee is composed of:

1. A representative of every carrier serving the airport, including non-ATA and foreign-flag carriers. These representatives are airline station managers or cargo service or passenger service managers. They are not security officers.
2. The airport manager or his designated representative, usually the airport chief of police.
3. A representative of each contract security company, catering service, baggage service, or other organization concerned with security of the airport.
4. A representative from every law-enforcement agency serving the airport including FBI, Customs, U.S. Postal Service, local police, and airport police.

5. A security representative of the Federal Aviation Administration (FAA). FAA tower personnel attend meetings of some of the committees.

A carrier security adviser for each committee is appointed by the parent ATA Security committee. This appointment is permanent unless change in duty assignment indicates the need to change advisers. The carrier security adviser makes every effort to have one of his security officers attend the monthly meetings (held at the same time and place). At the meetings and between meetings, the carrier security adviser is able to furnish professional security guidance and to suggest agenda items.

In keeping with the primary purpose of inculcating a security awareness in as many employees as possible, the officers, chairman, and secretary serve for one year.

The director of security of ATA serves as the focal point for local committee activity. He reviews the monthly minutes of the local committee meetings, sends out bulletins to keep chairmen advised of recent items of security interest, and advises all chairmen of any new idea or procedure that has been successfully used at another airport.

A good example was the "Meet and Greet" program that started in Charlotte, North Carolina. Most airports had challenge programs to intercept unbadged persons on the airside of the airport, and the local committee at Charlotte developed a card that was issued to all employees explaining the courteous but firm approach toward unbadged persons—to meet them and make sure that they were not trespassers.

Cargo-security items for consideration at the monthly meetings of the local committees include:

1. A report from the Cargo Security Task Force.
2. A report from the airport chief of police on airport crime in the previous month.
3. A report from the representative of the U.S. Postal Service on mail security at the airport, with suggestions on how it can be improved.
4. A report from the FBI representative on thefts from interstate shipments.

5. Any reports from teams doing security probes to ascertain how easy or difficult it is to enter cargo areas.
6. Discussions of problems associated with the identification-badge systems or with the vehicle-identification program.

It would be nice to say that all the local committees have been doing a great job, but that is not the case. An objective analysis of committee performance would reveal that 50 percent are doing excellent work, 30 percent good, 10 percent fair, and 10 percent poor.

ANALYSIS OF THEFT PATTERNS

A side benefit of having people in other disciplines chairing the local security committees is that a certain amount of irreverence toward traditional ways of doing things is helpful. The best qualification for being a task force member (chairman of a local security committee) seems to be not knowing all the ways in which things cannot be done.

An obvious benefit of having people meet on security matters every month is that each member becomes aware of what is happening to other tenants at the airport. Are any particular flight segments accounting for a high share of problems? Are contractors tailgating through the card-controlled vehicle-access gates? Are some carriers experiencing losses on certain shifts? Have there been any problems with drivers trying to bypass normal control procedures? These and many more items can be reviewed. The result is that carriers are able to take preventive measures to ensure that they will not become victims. In addition, law-enforcement personnel are more enthusiastic about working with a company that really cares about security.

Annual Workshops

Because chairmen of local ATA security committees change every year, the parent ATA Security Committee conducts an annual workshop for the new chairmen. Held every January, the workshop is designed to strengthen security awareness and help all concerned keep abreast of patterns of crime. Following are the agenda items from a recent conference:

- overview of the antihijacking program: delineation of responsibilities; cost effectiveness; roles and relationships of airlines, local committees, airports, government agencies, and contract security companies;

- what a contract security firm expects of an ATA security committee;
- what an airport police chief expects of an ATA security committee;
- a slide presentation on baggage security;
- crimes against flight attendants and cockpit crew;
- security of airport lockers;
- the FAA looks at aviation security;
- cargo security: the ten mortal enemies of cargo movements;
- fraudulent claims;
- chairing a local ATA security committee; and
- building security into construction and remodeling.

The Annual Directory

To ensure direct communication among all organizations concerned with aviation security, the security director of ATA publishes an annual directory of security officers of United States airlines. This includes U.S.-based security representatives of foreign carriers, chairmen of local ATA security committees, ATA fraud-prevention officers, Drug Enforcement Administration airport units, FAA headquarters and regional security personnel, FBI airport resident agents, U.S. Customs personnel of the Office of Investigations and Office of Inspection and Control, and postal inspectors at major airport mail facilities.

LIAISONS

Freight-Claims Prevention Seminar

The security directors of the member carriers of ATA work very closely with the freight-claims examiners of the airlines. The claims people have an annual freight-claims prevention seminar and dedicate a substantial portion of time to presentations of security officials. A security official once pointed out that a company with a two percent profit margin would have to generate sales of $50,000 to offset a $1,000 paid claim. Security officers have been able to call the attention of claims examiners to items that might appear in a claims file that could trigger a call to the security department. The claims examiner, in verifying the authenticity of a claim, is often able to point out to security or freight services personnel how the incident that led to the claim could have been prevented.

National Cargo Security Council (NCSC)

In the summer of 1971, the Transportation Association of America (TAA) and the U.S. Department of Transportation joined forces to present the first National Cargo Security Conference. Four hundred representatives of all modes of transportation, government officials, shippers, manufacturers, and labor leaders attended the conference, which was entitled, "Cargo Security Crisis."

Most speakers deplored the lack of security in the transportation industry and recited statistics by the yard. The airline-industry speaker said that there was a need for action instead of bemoaning past failures. He called on TAA to form a cargo security council that would be representative of all industry segments with an interest in cargo security.

The challenge was immediately accepted and the National Cargo Security Council was founded under the chairmanship of Harold Hammond, president of TAA. Mr. Hammond has since retired. The council serves as a forum for the exchange of ideas, and as a central liaison group with government counterparts. None of us is as smart as all of us. The current NCSC chairman is Lou Tyska and the executive director is Joe Baker.

International Associations

The International Air Transport Association (IATA) discontinued sponsoring an annual security and fraud-prevention seminar in the early 1970s. This left a void in full-scale international cooperation on security matters. Into the breach stepped the International Association of Airline Security Officers (IAASO). Under the leadership of North American and German carriers, the annual seminar of IAASO has become the most important of the aviation security seminars. Thirty-five countries from all continents and subcontinents were represented at the 1980 meeting. The carrier security officers all agree that if a cargo security problem develops in one corner of the world, it is comforting to know someone you can call for assistance. The principle that security is noncompetitive is becoming contagious. A recent incident illustrates this point.

I recently had occasion to seek some information on an explosive-resistant type of glass. I was advised that two foreign carriers had some experience with the product; one of the carriers was Lufthansa. The first carrier responded immediately with a short rundown on the product. Lufthansa, with typical thoroughness, sent pictures, test results, where the product was used—"the whole nine yards."

U.S. airline security directors have been among the leaders in the IATA Security Advisory Committee. They participate in surveys of international airports and have consistently chaired the Property Crimes Subcommittee dealing with prevention of cargo theft and pilferage.

American Society for Industrial Security (ASIS)

The security directors of the nation's air carriers have long recognized the advantages of close liaison with security executives in other modes of transportation and in other industries. They have been active members of the Transportation Council of ASIS.

Because many of the ASIS security directors reading this book work for companies that ship by air, they should be acquainted with two publications that will be of considerable help. The first covers packing guidelines for safe, secure delivery of goods and is titled *Role of Packing and Handling in Cargo Security.* It is available online from the DOT, publication No. FAA X-0228. The second is *Air Cargo from A to Z,* published by the Air Transport Association (ATA)—with packaging tips for air-freight shippers.

Government Liaison

The determination to do more as an industry and depend less on government has increased in the last few years. In a speech at the National Cargo Security Conference in Chicago in September 1973, Secretary of Transportation Dr. Claude S. Brinegar said:

> "The Federal arm is neither long enough nor clever enough to reach out from Washington to police the movement of goods every step of the way from origin to destination. That is not the Federal role. We will continue to be involved, but will look increasingly to the carriers,

> freight forwarders, wholesalers and warehousemen to adopt proven procedures for the protection of merchandise entrusted to them."

There are times, however, when preventive security breaks down because someone forgot to give advance notice of a high-value shipment, or someone failed to stow high-value cargo in the high-value cargo cage. In such cases, the airline will look for investigative assistance from the FBI, Customs, or the Postal Service.

Liaison is mutual. The FBI participants at local ATA security-committee meetings and national conferences and seminars are both donors and recipients of information. They can alert attendees to potential problem areas and they can receive information about patterns of crime that may be developing. In addition, there is the personal contact. The FBI agent will know whom he should see if he has an investigation at the airport.

The representatives of the U.S. Customs Service have been very active at many airports. They have presented cargo-security seminars for airline personnel and have made effective use of their patrol personnel. Through the cooperation of airline security and Customs officials, the revenue of the United States is protected. Airlines and Customs have been rewarded by a sharp decrease in theft of international air freight.

The theft of registered mail from airports sparked liaison between air-carrier personnel and postal inspectors. The postal inspectors seized the initiative by establishing the "con-con program," a system of *concentration* and *containerization* of mail. A third "con" could be added in that certain critical mail shipments were *convoyed* to and from planeside. The carrier security directors heeded the counsel of the postal inspectors and worked with them in developing sound procedures and security awareness, the inseparable ingredients of good security.

Security Awareness

It is difficult to appraise the increase in security awareness at the nation's airports, for that factor is an intangible. But in many airports you can sense it, almost touch it. You can see it in the police force, in the cargo area, at the ticket counter, on the ramp, at the screening points. Haste and boredom are missing from the environment, and there is a crisp confidence in the air.

This is not the case in a few airports. At these few airports, many of the people seem more concerned with their own interests, their own welfare. They fail to realize that the person pilfering something from a shipment, or the truck driver who managed to have twelve items loaded on his tuck and to sign for only eleven, is stealing their own jobs.

On balance, there seems to be a substantial increase in security awareness, as sound operational and security procedures are put into place. This is particularly evident when one reads the minutes of the monthly meetings of the local security committees and notices little items on potential security problems, items that probably would not have been noticed ten years ago or, even if noticed, not reported.

The Statistical Record

There is a commonly used measure of success or failure in protecting freight from damage or loss. It is the claims ratio—the amount of money paid in claims for every $100 of revenue. Claims are paid for three basic reasons: damage, delay, and theft. Theft claims include theft, pilferage, burglary, robbery, and any other shortage. All shortages are considered theft-related. It used to be that most air-freight claims were theft-related, but this is changing. In 1969 and 1970, for example, about 66 percent of claims were theft-related and 34 percent were from damage or delay. In the last few years, the percentage of theft-related claims have dropped to a little under 50 percent, whereas damage and delay claims have increased to a little over 50 percent.

A LOOK TO THE FUTURE

In our own lives we strive to become better men and women, yet we never become perfect. The same is true in security. The security officer is constantly trying to close the security circle, but there are always gaps or bubbles. The security officers of the member carriers of ATA must keep making the gaps and bubbles as small as possible. What gaps and bubbles are on the horizon?

Apathy. When things are going well, a false sense of comfort sets in and security awareness begins to evaporate. Security officers, individually and collectively, must ensure that they have a dynamic program with no room for boredom.

Liaison Between Personnel and Security Departments. This liaison should not be hit or miss; it must be constant and thorough.

Cooperation of Manufacturers. Such cooperation must include not only the normal safeguards such as good packaging and handling procedures. It must include a good identification system, particularly of merchandise attractive to a thief. It must include a record of all serial or identification numbers so that when thieves are arrested, manufacturers will be able to identify their products positively.

Selective Prosecution. The Justice Department policy in recent years has been the selective prosecution of crime. Five categories of crime receive special attention—they are the "in" crimes. The great majority of crimes receive precious little attention. The buzzword is that Federal authorities "defer to local prosecution." That sounds good, but it is not working. In many instances, local authorities decide to concentrate on the "in" crimes. The result is that much crime is untreated. One of the leading authorities on crime had this to say and his prophetic words ring more clearly every day: "Crime that is untreated will not fade away. Where selective prosecution is chosen as a social remedy over treating all crimes, positive alternatives must be found to deal with the problems left untreated, otherwise seeds of anarchy will find root and grow."

The Importance of Packing and Handling in Air Cargo Security

Cargo loss; cargo damage; claims; surveys and their cost; wasteful handling and its cost: all these things are instant financial drains eroding the profit in air cargo transportation. No airline should, for a moment, hesitate to take action that would materially reduce such avoidable waste of money, time, and customer goodwill. The report that follows provides for airline cargo, cargo sales, and security personnel an outline of the basic expertise in packing and handling necessary to protect carriers from unnecessary loss and damage situations.

PROPER PACKAGING IS AN INTEGRAL PART OF AIR CARGO SECURITY

Weight-Cost Factor in Packaging

Because in air-cargo transportation cost is a matter of weight, the use of light-weight containers is of vital concern to the shipper. What the shipper wants is as light a container as is consistent with reasonable protection of his cargo.

Problems in Selection of Packaging

As a general rule, shippers have insufficient knowledge of such fundamental packing techniques as container selection, protective interior packing, even at times of proper assembly and closing procedures. These shipping failures are further compounded by the fact that cargo handlers seldom can determine the cause of such basic failures as crushing, creasing, and tearing. These are serious deficiencies in handling and packing techniques, and the airlines must overcome them in order to protect themselves.

Defenses Against Poor Packaging

Various means are available to an airline to protect itself against and at least minimize the effects of poor packing. With this in mind, the following examines the problem of packing for air cargo, particularly in terms of the proper selection, packing, and handling of cartons.

Shipping Container Provides Only Limited Protection

The nature of the exterior carton alone is no guarantee that the merchandise will reach its final destination in usable condition. Unless goods are properly cushioned, blocked and braced within the container, the carton itself will provide inadequate protection for the contents. Be it also noted that the final closure, strapping, and marking of the container are equally vital in assuring the delivery of the merchandise undamaged.

Carton Materials

Almost universally, goods shipped by air are packed in light, highly adaptable fiberboard cartons. Since the large majority of U.S. shippers use fiberboard boxes that meet the requirements of the Uniform Freight Classification, there is every reason to conclude that the cartons, as such, are made of an adequate material. Why then the untenable amount of damage? Simply this: packing procedures—the carton is not of the proper size; the interior packing is inadequate; and the assembly and closure is not sound and secure.

Carton Selection

The proper selection of a carton for the shipping of a particular item involves four principal considerations:

1. Resistance to compression.
2. Strength of score lines (creased edges).
3. Resistance to puncture.
4. Ability of the fiberboard to resist the weakening effects of moisture.

Resistance to Compression

A carton's capacity to resist compression is, in general terms, inversely proportionate to the length of its corrugations. This principle holds, however, only when the carton is stacked with the corrugations vertical to the base upon which it rests. If the carton selected has corrugations too long to sustain the weight of the merchandise, or if the carton is placed on its side, crushing and the consequent opening of closures is inevitable.

Score Lines (Creased Edges)

If the closures of the carton remain secure, most pressure and breakage will occur along the score lines. Score-line joints vary considerably in their capacity to withstand pressure; however, it has been observed that foreign-made cartons, particularly those boxes in which the body joint is lapped and stitched, are the most likely to fail. Handlers should be most pointedly made aware of this fact.

Resistance to Puncture

Resistance to puncture varies with the materials and the workmanship of the finished box.

Resistance to Moisture

A plastic shroud poorly applied will expose the cargo to possible damage by rain. A plastic shroud can be properly applied by placing a sheet of plastic on the pallet board prior to the pallet buildup. After the pallet is full loaded, this bottom sheet can be drawn up to cover the lower portion of the load, and a second sheet can then be placed over the load to overlap the bottom sheet of plastic.

Types of Loads

As indicated above, the choice of a carton suitable to the character of the load is vital. The nature and importance of this matter of carton selection perhaps needs specification.

Loads are classified as (1) easy, (2) average, and (3) difficult in terms of the degree of support they give to the faces of the carton. Single items capable of completely supporting all faces of a carton are classified as *easy* loads. A shipment of radios, shoes in boxes, packaged hardware—all transmitting vigorous support to the faces of the carton—are characterized as *average* loads. Clothes, wigs, bolts of cloth, machine parts of varied and irregular shapes, because they obviously do not uniformly support the faces of the carton, are characteristically *difficult* loads. Patently then, a shipper who gives no consideration to the nature of the load in selecting a carton for its shipment, will frequently expose his cargo to unnecessary damage.

Load Bulge

As a supplemental problem, associated with load type and general carton quality, is *load bulge*, of which there are three groupings: filling, compression, and settling. Briefly (for the matter is rather obvious), an overloaded carton of loose wearing apparel is bulged by overstuffing; a carton with side bulges caused by overtightening of straps is bulge by compression; settling is self-evident. But whereas bulging weakens a carton, the greatest detriment to it comes from handling and from loose-unit loading in particular.

Load Collapse

The opposite of load bulge is the caved-in or collapsed load, which reveals itself in a crease or crush inward of at least one of the faces of the carton. Although container cave-in, like bulge, causes loose unit loads, its greater significance is that it evidences an oversize box for the load, inadequate interior packing, compression, or the having being hit. Care in the selection of the carton used reduces the risk of such carton failures.

Inadequate Interior Packaging

There are no published commercial specifications as to what constitutes adequate interior packing; therefore, how goods are blocked, braced, and cushioned within the carton is the responsibility of the shipper. They should be packed in such a way that they cannot move because any item that can move within a container invariably results in damage to itself, the container, or both.

Improper Closure

Improper closure occurs when the top and bottom flaps are not tight together and square. This results from poor or inadequate application of closure devices, such as staples, adhesive, or tape. This failure occurs because of an oversized box, allowing the contents to shift within the carton.

Indicators of Weakness

The most obvious indicator of a weakness in packing is that the type or nature of the trouble will be common to more than one package. This weakness is particularly easy to spot in unit loads, where all packages of the unit are alike, because all weaknesses are in the same location in several or all boxes, and all evidence one or more of the packing failures discussed above.

PROBLEMS COMMON TO SHIPPERS AND CARRIERS

Forces

Damage results from hazardous forces encountered in transportation, handling, and storage. Whether the fault of shipper or carrier, damage results from abrupt starts and stops, or from vibrations and jolts. Handling forces are those received through loading, unloading, and handling during storage operations. Damage often occurs during manual handling, forklift truck handling, and cargo loading, as a result of dropping, puncture, crushing, wracking, and smashing of the containers. Storage forces are those resulting from the crushing effects of loads superimposed through too high or too heavy stacking.

Exposure

Exposure to different climatic conditions and weather hazards such as high humidity, rain, extreme cold, dry intense heat, and the cycling of these weather conditions tend to accelerate the breakdown or deterioration of unprotected items. Water in its various forms is the main damaging factor in climatic exposure.

How Packing Helps

A summation of this commentary actually can be sharply focused. Items that are packed properly will, for the most part, resist the damaging effects of force and exposure. Force is counteracted by selecting the right shipping containers; immobilizing the item within the container through blocking and bracing; absorbing forces through the use of cushioning material; and properly assembling and closing containers. Exposure is counteracted by using water-resistant shipping containers and waterproof barrier materials in their various applications. Storage under cover at all times is the most effective means of resisting exposure.

WHAT AIRLINES CAN DO

Although it is true that an airline can bring many of the weaknesses of packaging to the attention of shippers through sales promotion and general education programs, with a view to improving shipper practices, the carrier's best protection against such packing deficiencies is in recognizing the problems and in taking appropriate measures when problems are discovered. These protective actions should be taken consistently and comprehensively.

Examine Packing

Any packing which is hopelessly inadequate should be refused for air carriage. Where the packing can be put into shippable condition for transportation, require the shipper to do so, making notations on the waybill of what action was taken to make the package acceptable. The carrier must be mindful of the fact that the acceptance of such cargo for shipment in all probability will involve special handling of some sort, with the additional cost connected with such handling. In addition, the carrier cannot disregard the fact that poorly packaged cargo increases the potential of damage claims at destination.

American-made cartons usually show the gross weight limit in the Manufacturers Certificate on the bottom of the carton. Weighing the carton and comparing its weight with that of the certificate will show if the container is overweight and, therefore, improperly packed.

Stow Cartons Upright

At the risk of some repetition, it seems proper, at this point, to call attention again to all the precautions made explicit in the section on Resistance to Compression with respect to storage and stowage of cartons.

Keep Pallets in Good Repair

Pallets with broken or missing top and bottom slats, or exposed and protruding end stringers, cause damage. Such pallets should be immediately repaired or removed from service.

Build Unit Loads Properly

In building up a pallet board or loading a container, heavy items should be stowed on the bottom and light packages at the top. When cartons are placed on pallets or skids, the bottoms should not extend off the edge of the pallet; otherwise, the overhanging portions will crease and weaken the cartons. When cargo is *stacked, maximum compression strength from each box is* obtained if all packages are kept in precise vertical alignment. If this is not done, the result is the same as loose loading or allowing packages to hang over the edge of pallets or skids.

Protect Packages From Netting on Pallet Boards

Extreme care should be taken once loads on pallet boards are netted down. Under such conditions great forces are exerted on the packages at the top edges of the load. Relief from some of the net pressure is made possible by the use of edge protectors, made from folded fiberboard pads, or lumber in the form of a right angle, or the use of dunnage bags.

The best system for loading pallets with uniform size packages is to use one of the standard pallet patterns. The loading patterns interlock the packages and provide greater distribution of the weight to all areas of the load, resulting in a tight load.

Use Lift Trucks Properly

The proper buildup of pallets, operation of lift trucks, type of lift truck used, and upkeep of such equipment are critical in preventing damage to cartons through creasing, puncturing, gouging, tearing, and crushing. These problems occur at shippers' and consignees' facilities as well as in carrier terminals.

THESE THINGS MUST BE DONE

Airlines must *protect themselves against poor packaging* by shippers, and they themselves must *handle packages properly.*

Packing, particularly packing involving the use of the carton, was not a major factor when cargo was stuffed indifferently into the belly compartment of a relatively small aircraft. But with the advent of the jet freighter, stowage and handling required change. Practices in cargo handling suitable in the stuffing of belly compartments no longer are applicable. Yet the fault is that precisely these methods have lingered on, increasing the incidence of loss and damage. Because there have been no significant changes in the carton as an exterior shipping container (nor are there likely to be in the immediate future), it is obvious that the carriers must urge, influence, and pressure shippers into proper utilization of available containers; and obviously the airlines must handle cartons with proper regard to their protection.

None of this is revolutionary. These suggestions do not call for new equipment or new "know how." The challenge here is to be alert to the actions and conditions that affect each day's cargo movements and to increase efficiency.

Better packing and better handling mean a better record for safe delivery. They also mean an important boost in the quality of air-cargo security through reduced opportunities for theft and pilferage.

Airports—The Battle to Deter Hijackers

Part One: Beating the System

Saturday evening was warm and a gentle breeze stirred the fronds of tall palm trees in the forecourt of the airport departures building. It was a special day for Akira, who had arrived very early to join his flight to London, because he was about to embark upon his first operation for the terrorist Red Brigade.

For two years Akira had been a staunch supporter among the 1,000 sympathizers mostly drawn from the elite universities. But now he was about to move into an upper echelon, although he was unlikely ever to know anyone other than his companion, Kazuo, who was to share the operation with him and, of course, his commander.

Akira had graduated from jungle-based guerrilla training school. Several weeks of arduous and intensive training at the hands of techniques, weaponry and the finer points of urban guerrilla warfare—such as the value of detailed reconnaissance and the dividends paid for patient and meticulous observation.

THE TERRORIST PLAN

Every day for three weeks, Akira and Kazuo mingled with the airport crowds, but never together. They wandered around the departures building, each carrying a lightweight cabin bag and seeming to be average Japanese tourists. Both men dressed differently for each visit, so far as a necessarily limited wardrobe permitted, and such was their attention to detail that even their spectacle frames appeared to vary from visit to visit, ranging from steel to gold and horn-rimmed.

As a result, both Akira and Kazuo were as familiar with the layout of the public areas of the building as their own homes. They noted the procedures used by the airline staff, and further, that these procedures never varied—although the *attitude of the staff* at the desks varied considerably, as did the length of their coffee breaks.

Armed with blank boarding cards easily stolen from an unattended check-in desk and empty coupon books surreptitiously removed from a rubbish bin in the arrivals building concourse, they met in the bar of their downtown hotel to compare notes and found no significant differences.

It was agreed that on the Thursday, Akira alone should make his way past a police guard at the entrance to the domestic departures lounge from whence, through a decorative but clear glass partition separating it from international passengers, he could observe the processing and security screening procedures. He would write his notes on paper concealed in a magazine.

On Friday, Kazuo would repeat the process and then the notes would be compared. In fact, their notes agreed entirely. Beside each air-bridge door a seating area, enclosed only by a decorative handrail on chrome supports about one meter high, was provided for the comfort of departing passengers. A baggage x-ray machine and metal detecting arch stood just inside the entrance to the enclosure, whilst at the far end, close to the air-bridge entrance, the exit gap in the barrier rail was closed by a simple, maroon-colored rope to which was fastened a chromium hook.

The majority of passengers drifted into the seating area well in advance of the security screening staff, who always arrived at the latest possible moment—due no doubt to an over-full schedule in processing other flights. The passengers were then asked to return to the entrance for screening. The keen travelers would leap to their feet to be first in line. Whilst the elderly, lazy, or simply experienced passengers remained seated, contented to await a more leisurely walk through the controls.

Thus Akira saw that several dozen screened passengers returned to their original seats whilst many unscreened people remained seated. "No control at all," thought the amazed Akira, "just sit tight and let it all happen, for the authorities cannot tell who has been through the control procedure and who has not."

His attention was then drawn to a small attaché case lying under the seat opposite him where a particularly attractive girl had been sitting, and his heart gave an extra beat. It would indeed be ironic if he should become the victim of a sabotage attack. He looked around quickly and was relieved to see her dutifully standing in the line to pass through the magnetometer arch, but without her bag, which, consequently, would not be passed through the security check.

Akira looked around with renewed interest. A passenger who arrived after the security screening had commenced could slide a bag under the handrail and then safely go to the check-point, after which he would simply occupy the nearest seat to his bag and await the boarding call. If, unexpectedly, a guard should be patrolling the lounge, he decided that the best method would be to leave the bag beside one of the seats backing on to the handrail and retrieve it surreptitiously by pulling it under the rail.

Later that day, he outlined the weaknesses for Kazuo, who confirmed them on Friday when they discussed their final plan for hijacking.

THE ATTACK

Thus it was that Akira arrived at the airport early on the Saturday, followed ten minutes later by Kazuo. Each carried a revolver and a grenade in his cabin bag and a nondescript suitcase containing the usual miscellany of personal gear.

Akira shuffled forward in the inevitable queue moving slowly towards the check-in counter. He gave not the slightest flicker of recognition as he saw Kazuo join the end of the line; indeed, fully confident of success, he sauntered up the staircase to the airside concourse and into the Gate 2 lounge, as he had been instructed to by the airline employee at the check-in counter.

The lounge was large, with seating for perhaps 300 people, and at this early stage was sparsely occupied. Although all of the passengers had been allocated the seat of their choice, nevertheless, the more experienced travelers had moved to the end of the lounge to seats nearest the exit, as though afraid the flight would leave without them. Unlike Kazuo, Akira neither drank nor smoked, so ignoring the duty free shop, he walked past the unoccupied security checkpoint and selected a seat backing on to the hand-rail barrier separating the lounge from the public concourse. This position, he thought, was doubly safe from observation should the guards, when they arrived, be above average in their vigilance.

From this position he could see neither the Gate 2 check-point nor, should there be a departing flight from Gate 3, the guards who would be screening that flight. The line of sight was effectively blocked by the roof support columns. Furthermore, if for some unforeseen reason Kazuo needed to pass an urgent comment, he could do so from the concourse whilst bending down to "re-tie a shoelace."

Indeed he did attend to a shoelace, but only to push his bag surreptitiously under a seat further along the row. Kazuo then moved away to the duty free shop where, after a leisurely inspection, he eventually made his purchases—by which time the security staff had arrived and the passengers were already spilling out of the lounge to form a line in front of the x-ray machine.

As the terrorists had anticipated, some people remained seated in the lounge and it was impossible to determine who were the already screened passengers and who had not yet bothered to move. Kazuo was among the latter group—but with a slight difference: he did not intend to move until the final call-forward was made.

The subsequent hijack became another statistic in the record of infamy. By some authorities, it was listed as "successful" because the aircraft captain had been unlawfully relieved of his command. But to others, who should know better, it was classed as "unsuccessful" because, during a refueling stop, the aircraft was stormed by soldiers and the hijack terminated.

It's true that neither Akira nor Kazuo achieved their objective—in fact, Kazuo died instantly with a bullet lodged in his brain whilst Akira is still serving a ten-year prison sentence. Nevertheless, everything that happens in the air has its origins on the ground—and the ground defenses had successfully been beaten.

THE SECURITY PROBLEM

Life is full of "ifs," "buts," and wisdom gained with hindsight. If only a floor-to-ceiling glazed screen had been installed to separate the gate lounge from the concourse instead of open hand-rails, the spacious open effect would have been maintained and at least two methods of evasion of control would have been prevented.

The other essentials to firmly close the remaining gaps would be to provide additional seating in the public concourse and, by closing off the lounge with glass doors, ensure that passengers could not enter the lounge until the security staff arrived to unlock the door and commence the security search.

The naïve rope barrier at the exit from the lounge should also have been replaced by a sliding glass door, to be unlocked by airline staff when they were ready to receive passengers for boarding card check at the entrance to the air-bridge.

Commonsense and imagination—plus experience—have a major role to play in security precautions. If the security officer has also been blessed with a criminal type mind, but of course has not himself succumbed and exceeded the bounds of propriety, he is well along the road to becoming an effective officer.

THERE'S MANY A SLIP

Several years ago I was responsible for designing and implementing ground defenses at an airport where stringent security precautions were necessary.

An x-ray machine, suitable for accepting cargo and, therefore, the largest pieces of hold baggage as well as cabin baggage, was placed just inside the entrance to the terminal building. Everything hand-carried by passengers and staff was examined by x-ray before it was allowed to be taken further into the building. Later in the process, and after the passenger had been relieved of his hold baggage at check-in, he was physically searched by security guards before boarding. *If you think this system was foolproof, you're wrong.*

One day, a man placed his suitcase and briefcase into the x-ray machine and the contents proved to be quite innocent. He then moved across the concourse to a check-in counter where his baggage items were weighed, labeled, and placed on the baggage belt.

Just as his briefcase was about to be whisked away, he asked for it back, making the excuse that he had forgotten to take out some papers needed on the flight. He knelt down—and with his back to the security staff operating the x-ray machine, took papers from the brief-case, replacing them with an automatic pistol taken from inside his jacket.

The briefcase was then handed back to the counter clerk. He went through the body search check-point without any difficulty, of course, and joined his fellow travelers on the apron where the hold baggage was waiting on a truck to be loaded.

Walking across to the loader, he asked if he might be allowed to take some papers, which he "needed during the flight," from his briefcase. The helpful loader readily agreed, but only glanced at the man whilst the case was unlocked, and the removal of the gun to the jacket pocket was not seen.

Thus, a gun was taken into the passenger cabin and my supposedly immaculate defenses had been beaten. It will not happen again at that airport, nor, I hope, or at any airport where the staff have been properly trained.

EQUIPMENT

Necessity and war may both be mothers of invention. Certainly, in the war against aviation terrorism, technology in protective devices is constantly moving forward at great speed.

The true incident related above could have been avoided in several ways. First, there might have been metal detecting devices beside the x-ray equipment at the entrance. This, alone, would have prevented the pistol from being carried into the building and denied an opportunity for switching it between pocket and bag and back again.

If the bags, when they emerged from the x-ray machine, had been sealed with polypropylene strapping—as done on Concorde flights—it would have been impossible to add or extract anything from the case without breaking the seal, and the breaking strain is 270 pounds!

If the insertion of new items after a security search is prevented, likewise it's not possible to remove items either. *The prevention of pilfering from bags thus achieved is an added bonus.* Indeed, airline carriers should weigh the capital cost of strapping every bag against the annual cost of settling passengers' claims (many thought to be fraudulent) for compensation following "loss."

Neither the employee at check-in nor the loader should have returned a registered bag, except in the most exceptional circumstances. Exceptional and compelling reasons are, however, as plentiful as tabby cats and range from passport, ticket, and money being left in a bag to the frightening and unquestionable need to retrieve a medicine, the prolonged absence of which threatens imminent death—or so it is often claimed. In these circumstances, the item may be taken from the *bag under supervision.* If the bag had previously been sealed, it will have to be resealed for it should have been firmly established when bag sealing was introduced, that unsealed bags would be accepted neither by check-in staff nor the loaders.

Just as witnessing a bad traffic accident will cause most of us to drive cautiously for a few miles until the horror and shock fades from our memory, so a security disaster at an airport provides extra security awareness and vigilance for two or even three weeks following the disaster.

Admittedly, much security searching becomes a boring and routine chore, especially at a domestic airport, and herein lies the risk of guards becoming complacent. Yet an astonishing collection of actual or potentially dangerous items are found in baggage every day at a major international airport.

When the security searching of passenger baggage began in earnest in the late 1960s at London's Heathrow Airport, the daily haul averaged 50 items. In recent times a press report claimed that in one Heathrow terminal alone the haul for one day was five bayonets; three chemistry sets; two shotguns; a Gurka kukri; a crossbow; a swordstick; a sword; a knife; a toy rifle; seven toy guns; an axe; and a spade. Some of these items were probably souvenirs—but why would a passenger take an axe and a spade into an aircraft cabin?

It is most heartening that the security guards at Heathrow have made it the safest airport in the world, and much the same can be said for the remaining British Airports Authority airports. *But in case any of them should read these words, let them not consider themselves perfect.* A terrorist or criminal lunatic starts with the advantage of a cunning, devious, and warped mind. Furthermore, he has all the time in the world to observe and plan.

One of my early instructors had a quaint expression: "Folk will persist in being people." An American security colleague of mine puts it another way and simply says: "We have people problems." Security staff are mostly nice, normal people who are reasonably law-abiding like the rest of us. It's expecting too much that they should anticipate every move the enemies of civil aviation make, and for this reason, we can never ensure total success. However, we can learn from each experience and ensure that we do not make the same mistakes twice.

DEVIOUS AND CUNNING

I have a family of toy animals designed and made for me by my clever and patient wife. The animals' stomachs are lined and shaped to conceal a most realistic automatic pistol made of gun metal, an ounce of simulated drugs, and a round of ammunition. The most traveled animal is a dog, code name "Murgatroyd," who, when engaged upon operational training, is enclosed in a clear plastic bag labeled: "To Mary with love from Uncle. . .", and he travels either openly under my arm or in my cabin bag, depending upon the particular security screening arrangement and thoroughness of the security check.

On one occasion, tense and anxious even though the local airport manager and police knew of my presence, I could see from the lounge that a reasonable check was being made of bodies and baggage at the departure gates. It seemed likely I would have to abandon the attempt to evade the controls as I watched the exclusively female staff placing bags in an x-ray machine and marshalling the passengers of an earlier flight through a metal detecting arch.

It was then I noticed a couple with a little girl, whom I judged to be about six years old, taking coffee at a table in the lounge. Arming myself with a bar of chocolate, I sat near them and took Murgatroyd from his plastic bag. Very soon the child's attention was fixed upon the dog and a short session of peek-a-boo with the animal round the table leg raised a chuckle and brought her to my table.

Her parents watched with pleasure my seduction of their little daughter. Our short chat together remains forever a secret, but I did finally ask her if she would like to carry my doggy to the "aero plane"—a suggestion which met with her unqualified approval. I repeated the dog's name—July—several times and asked her if she liked it and could she say it. Yes...she could say it, very well indeed.

I placed myself close behind the family in the queue so that if matters took a wrong turn, I could prevent a scene and any embarrassment. Nothing, however, did go wrong.

As the child was gently pushed forward by her mother, as daddy put the cabin bags on the x-ray belt, a security woman at the arch said: "Hello, darling, what a lovely doggy. What's his name?" at the same time taking the toy from the child's hand. The child shyly passed between the x-ray machine and arch whilst the security guard crouched down to coax an answer from her.

We were safe. The defenses had been beaten. The handling over of a bar of chocolate in exchange for the dog, coupled with a few fond farewells, took only a matter of minutes. Whilst I am not particularly proud of using a small child in this way, the point had to be made, and the family was not exposed to any kind of danger.

The lesson is very simple, however. If the procedure requires every living thing to walk through a detector and everything carried by hand to be placed in an x-ray machine (or alternatively be given a realistic physical search) then this *must* be done. There must be no exceptions made for attractive children and cuddly toys. We cannot afford to take things on trust. The familiar everyday sight is especially useful to the terrorist because we tend not to “see” the commonplace things.

CHEAP—BUT EFFECTIVE

I was very concerned about a post office van standing beside an aircraft being loaded with sacks of mail. There was no easy or speedy way of checking that the van on the apron was the one which had left the post office on an authorized journey, nor indeed that the driver had not stopped at his favorite inn for a drink before arriving at the airport, thus providing an opportunity for a bomb to be put into the vehicle.

With the ready cooperation of the postmaster in the city and police at the airport, the solution to this problem was cheap, simple, and most effective. A supply of serially numbered colored plastic seals was purchased. A different color, within a limited range, was used in a random pattern for each journey. The sliding door behind the driver was sealed by passing the plastic strap through two metal loops welded to the metal door, and the rear door was treated in the same way. The only way to gain access to the interior of the van was to break one or other of the seals, which could not then possibly be resealed or re-used.

Before he commenced his journey to the airport, the driver was handed two identical envelopes—one addressed to the police at the airside access gate, the other to the airline carrier. The van was loaded with mail bags and sealed within post office premises. Upon the van’s arrival at the airport gate, the police opened their envelope and read details of the driver’s identity card number, noted his name and were able to see a colored photograph of the holder. Also enclosed were details of the color of the seals and their respective serial numbers.

When these details had been physically checked and the bottom of the van had been examined with a mirror mounted on a trolley to detect the presence of any bomb attached to the underside, the vehicle was allowed to go on to the apron with the seals intact. Upon arrival alongside the aircraft, a senior airline official opened his envelope and ensured, on behalf of his airline, that the seals were of the correct color, serial number, and were intact. He then cut the seals and put them in his pocket to support a subsequent entry in his logbook. He did not, of course, need to check the driver's I.D. card nor make a physical search under the vehicle for a planted bomb—the police always did their check zealously.

CONCLUSION

Machines unaided do not commit crime. On the credit side they do not tire nor sneak off for coffee, and they are unable to argue or use discretion. But machines, however sophisticated, are only tools and cannot perform the total security task. It is the flexible human brain, which will interpret a complex situation and make a decision. The fallible human being must be boss—but his fallibility is more likely to be reduced if he holds a sincere belief in the rightness of his task—the saving of life and the prevention of crime.

Common sense and a refusal to over-react to incidents—or to hysterical and ill-informed publicity—has a major role to play in designing and implementing security systems. Additional qualities, however, are also required, not only in the specialized area of aviation security, but throughout the world of security generally. These qualities are the "CID" of the dedicated security operative and include:

- *Courage* to undertake an unpalatable but essential course of action and to recognize the occasion when a calculated risk may be taken;
- *Imagination* to see potential risks and assess the depth of action necessary—yet not impair the operational functions of the industry; and
- *Determination*, within the law, to bring to fruition measures designed to circumvent the threat.

Part Two:

The Successful Hijack

One of the principal aims of an aviation security program is to prevent the commission of any criminal act against an aircraft. But it's also necessary to consider what should be done if the security system is breached and an individual or group succeeds in seizing control of an aircraft. In the civil aviation industry, tried and proved emergency procedures have been drawn up, practiced, and implemented to deal with all possible emergencies which might arise during a flight. But it is doubtful if any of them have presented the unique problems which must be considered before recommending the best way to handle a hijack.

It is now generally agreed by the majority of carriers, aviation organizations, and responsible authorities that the first and most important objective once a hijack has occurred is to preserve the safety of passengers and crew. And it's in accordance with this principle that all subsequent advice is now being formulated.

It is prudent that the reader should be informed at this point that the remainder of this article does *not* disclose a detailed set of procedures for handling a hijack situation. The reasons for this are: (1) They do not exist; and (2) If they did, there would be no proper way in which they could be given unrestricted dissemination.

HIGH STAKES

A hijack can be compared to a game of cards—bridge, for example. No one can tell which individual cards will be dealt to each player, neither can the order or way in which each player will use his cards be determined. It *is* possible, however, to teach players the principles of the game, so that by intelligent application of those principles they can make the best possible use of the cards dealt to them—and make a proper response to a partner's initiative. Partners with a good knowledge of the principles of the game will have a better chance of winning the game, or at least minimizing any losses.

When we remind ourselves that the stakes in the game of hijack are human lives, the wisdom of training one's players becomes indisputable. Perhaps it's best at this stage to dispel any misconceptions that might exist concerning a secret method or weapon which can be used to overcome any hijack attempt. If such a method or weapon has not yet been devised, it is not due to the lack of well-meaning advice, which is constantly offered to aviation security officials!

Starting with the original idea years ago of traveling sky marshals, other ideas put forward vary from sophisticated remote surveillance systems for installation in aircraft; total isolation of the flight deck from the rest of the aircraft; and gas dispensers controlled by the captain to anaesthetize at a touch the occupants of the passenger cabins. All of these proposals are unacceptable for the same reason: they are not in accordance with the principle of ensuring first priority for the safety of passengers and crew.

Any shoot-out with in-flight security, despite what may be said about the use of special weapons, creates a highly hazardous situation. Not only do we have to calculate the risk of triggering overreaction by a stressed hijacker who has a heavy caliber weapon or grenade, but we cannot overrule the strong possibility of the sky marshal presenting an easy target for any member of the terrorist group who has not yet revealed himself or herself.

Before reaching any conclusions about the best way of dealing with a hijack, it's also necessary to examine the motives of the hijacker. A hijacking is not usually an act of attrition mounted with the intention of destroying the aircraft and killing all on board. Such acts of terrorism do occur, but usually take the form of a direct attack on parked aircraft, or on an air terminal—or the act of aggression takes the form of sabotage.

MOTIVES

A hijacking is in essence an attempt to hold an aircraft and its occupants as hostages to ensure compliance with the hijackers demands. Those demands may be actuated by political or criminal motives and will most likely result in prolonged negotiations, so it's reasonable to assume that unless something triggers precipitate action, the hijackers will not wish to destroy their most important asset—the hostages.

But we must also realize that while the above reasoning can be applied with some certainty to hijacks mounted by experienced terrorists or criminals, the situation can be drastically different when dealing with a mentally-unbalanced individual or an enthusiastic amateur, who is likely to panic and act without reasoning.

CREW GUIDELINES

To some, it may seem that the situation permits a policy of when the hijack is inevitable, lie back and hope you come out with the least harm possible. This does not have to be the case.

Although it is not possible to draw up a set of specific actions to take in dealing with any type of hijack, it is possible to draw up guidelines which will enable crew members to make an informed assessment of their particular situation and assist them in deciding on the proper action to take. Training in the guidelines will also create confidence and help remove the fear that a decision has been based on faulty reasoning. And this confidence will also sustain morale during the long and difficult period ahead.

In formulating guidelines for crew members, it's necessary to cover the following aspects:

- Maintenance of the principle that the safety of passengers and crew must come first.
- The need to avoid triggering a panic reaction on the part of a hijacker—especially during the highly emotional first moments of an incident—must be stressed.
- The best way to establish and maintain a relationship between the hijackers and the crew, which, while not amounting to total surrender and cooperation, clearly establishes in the hijacker's mind the need for preserving the crew's operational skills.

The advice for dealing with an organized group of terrorists will obviously be different when dealing with a single individual who is probably mentally unbalanced. But the risks in dealing with such an individual must not be underrated. Considerable care and tact is required. Care must be taken to avoid any triggering of a situation by inadvertently alarming the individual, arousing anger by a disparaging or inflammatory remark, or failure to give proper warning of some change in the situation (e.g., sudden aircraft noise, descent, or unusual activity).

An understanding of the likely course and nature of the prolonged negotiations which will take place helps to ensure that morale is not shattered by the inevitable frustrations, delays, and deadlines which will occur. Advice on the care and attention of passengers to safely bring them through a long and difficult situation should not only include physical requirements, such as the organization of meals, hygiene, toilet arrangements, and care for those with medical problems, but also cover maintenance of morale and the dissuading of any well-meaning, but ill-informed, person from initiating individual action against the hijackers.

THE PASSENGERS

It's natural that at some time during a hijack a member of the crew or a passenger could well be tempted to "have a go." Guidelines should cover this point. In dealing with a well-organized terrorist group, their vigilance, numbers, and weapons will certainly make it apparent to even the most rash individual that such action could only result in disaster. There is probably a greater temptation to "have a go" when dealing with the "loner," especially when there is cause to believe that the weapon or device displayed is only a replica or toy. On some occasions, the weapon might not even be openly displayed. In such cases the first principle of safety is still paramount.

Before any action is attempted, the situation should first be discussed; observations of the crew compared, and advice sought (if unrestricted communication with ground authorities is possible). Usually, the safest course is to wait until the aircraft has landed safely and let the experts on the ground deal with the situation.

Guidance must also be given concerning the responsibilities of the captain and individual crew members during a hijack. This should include a clear indication of the policy which the captain is expected to adopt in the event of being able to obtain the release of certain individuals—including himself—from the aircraft, and the part which he should play in any negotiations or discussion of demands with the hijacker.

ON THE GROUND

Thus far we have only considered a hijack from the point of view of those in the aircraft. We have established that our intention is to get the aircraft down safely and then maintain a posture that will enable the negotiation phase to be successfully implemented. In achieving our objective, we also have to consider the problems and likely attitudes of the government and aviation authorities who will also become involved.

A hijacker must eventually decide where he wishes the aircraft to land. A landing may be for the purposes of refueling and necessary servicing to enable the hijacker to continue the flight to the destination where he ultimately wishes the final negotiating phase to be played out.

In past incidents, some airports have shown a marked reluctance to permit a hijacked aircraft to land, even temporarily. Airports have been declared closed, lights extinguished, or runways actually physically blocked. The International Air Transport Association (IATA) and the International Airline Pilots Association (IAPA) have made their views quite clear on this point: *a hijacked aircraft should be treated as an aircraft in distress and should be given every priority for a safe landing if it requests this.* It is hoped that through the medium of International Civil Aviation Organization (ICAO) all states will accept a policy that does not unnecessarily hazard an aircraft. Most have already.

An airport security plan will almost certainly include the nomination of an official to control the incident. This ground controller will probably be a police officer or military officer, depending on the local law enforcement organization. But whoever he is, he will be faced with an arduous and unenviable task.

It is, therefore, to be hoped that he will have drawn up a contingency plan based on all the expert advice available. If he has been well advised, he will already have covered the following important aspects of his plan:

1. Firstly, the location and equipping of his command post, which will need good communications to and from all units and personnel deployed under his command, as well as good communications back to his superiors and government officials. The importance of this rear link will be dealt with later.

2. When nominating the staff to man the command post, the commander should have considered which expert advice he will need to have readily available. It is important to include technical experts in the operation and characteristics of the hijacked aircraft. Experts from the airline involved can often provide this technical assistance.

3. Finally, the commander will need the services of a trained negotiating team and possibly interpreters. It will also be necessary to support the negotiating team by establishing the best possible communication with the aircraft. Again, the technical advice of the airline experts can be invaluable in this respect.

It is also to be hoped the commander is aware of, and dedicated to, the principle that the prime objective is to preserve the safety of the hostages. It is hoped that aided by his team of experts, he will be in the best possible position to negotiate through the situation and eventually persuade the hijackers to give up peacefully.

THE PRESSURES

The commander, however, can find himself knowing full well which course of action he would wish to take, but being subjected to pressures from his own superiors, brought about by political complications.

The possible international and political complications can be many. For example, an incident involving an aircraft of a foreign state may include hostages of many nationalities, held by hijackers of yet another nationality, placing demands possibly involving concessions, or release of prisoners by yet another state. While an agreed response is being formulated at inter-governmental level, the unfortunate situation commander must be able to maintain the situation on the ground and successfully negotiate from one deadline to another.

The nightmare which haunts every commander during a hijack is the moment when it becomes apparent that negotiation is no longer possible. This situation will occur when the hijackers start to kill hostages, or when it becomes certain that they will implement such a threat once the next deadline is reached. It might also occur if higher authority decides that political considerations prevent any further negotiation. Such a decision is a heavy responsibility, which in turn devolves ever greater responsibility on the commander selected to put an end to the situation in a way which will ensure the best possible chances of preserving the safety of the hostages.

In such a situation the need for coordination, liaison, consideration of technical advice, and above all, the training of all involved in the operation, becomes paramount. The balance between complete success and disastrous failure is balanced on such a fine edge that no factor on the positive side can be neglected or taken for granted.

The hijacking that occurred in Mogadishu in April 1998, showed what can be achieved when all of these essential factors are incorporated in the final action. In that incident the hostages included one U.S. citizen, one German, one Belgium, one Frenchman, one Norwegian, and two Swiss. The two pilots, a South African and a Kenyan, also were held. The hijackers demanded 100,000 ransom. The hostages were released unharmed after the international community pressured the hijackers' leaders. The hijackers did not receive any ransom.

HIGH RISK

What of the future? Statistics show that although ground security procedures are containing the incidence of hijacking, a number are still occurring. One is too many. The general threat to civil aviation must be assessed as high risk.

So long as there are frustrated groups or individuals with real or imagined political or personal grievances; those who are prepared to resort to violence for criminal purposes; and the mentally unstable, we cannot ignore the fact that civil aviation presents an attractive target.

What then can be done to provide the ultimate deterrent? There can be no doubt that a prime consideration to any hijacker is where to find asylum after the successful commission of the crime and when his demands have been met. *The whole operation is pointless if this asylum is denied.* An increasing number of states are recognizing hijacking as an international crime and have agreed either to bring the hijacker to trial themselves, or to extradite terrorists for trial by states against whose laws an offense has been committed.

While it is encouraging to note the way in which countries such as Cuba are returning hijackers, and to respect the decision of some European countries to put on trial political refugees who have committed hijacking as a means of escape, this does not by any means indicate that full international cooperation can be relied upon in such situations. Proposals have been made concerning sanctions, especially measures restricting civil aviation facilities against states guilty of providing asylum to a hijacker. Unfortunately, political, economic, and commercial considerations can drastically undermine the will to implement such sanctions.

Many advocate the mandatory imposition of the death penalty for such crimes. It must be accepted that this would minimize the increased risks which a state and all who operate under its national flag incur when members of a terrorist group are held in prison. Such a situation can trigger further hijack attempts to obtain their release. However, it is doubtful if world opinion and political pressures would permit the universal imposition of summary execution for the offense.

Successful operations to terminate hijacks by assaulting the aircraft have given rise to speculation that a rigid refusal to negotiate, combined with the certainty that force will be used to terminate the incident, could be the best deterrent. Again, one must consider the international will and accord of all states to implement this *and to continue to implement it resolutely.* Despite the inevitable incident which results in a major loss of life, the will to implement such a risky policy by a country where citizens of another country are primarily involved, also reduces the resolution to pursue such a policy far enough to make it bite.

The key to the future lies in reviewing what we have already achieved, and seeing how this can best be developed. The hijacking problem *has* been contained. We have achieved this by international cooperation in implementing proper security standards at international airports.

But we have not yet reduced hijacking incidents to the lowest attainable minimum. The reasons for this are that there are still loopholes in the security systems which must protect our whole system of civil aviation and its network of routes and airports.

While we continue with our efforts in this direction, which must be resolutely pursued in the face of rising costs, changing methods, new passenger products, and non-security designed airports, we must also ensure that in the event of a hijacker slipping into the system, everyone knows the best way to deal with the situation and has been trained and equipped to do what is necessary.

Part Three

SEAPORT SECURITY

SEAPORT SECURITY

MARINE CARGO SECURITY

These comments on marine cargo security are the product of efforts by the Port Security Committee of the American Association of Port Authorities (AAPA). It is designed to set forth the principles of port security for port managers and terminal operators.

The objective of port security is the prevention of cargo loss. This can be achieved by establishing and maintaining an environment hostile to theft and pilferage. *Standards for Marine Cargo Security*, first published by AAPA in 1980 and currently under revision, defines and analyzes the responsibilities for cargo security; describes procedural security, including cargo flow and documentation practices; lists areas of vulnerability, and prescribes standards for physical security. Training programs for security personnel are outlined in the book.

THE PILFERAGE PROBLEM

Pilferage is probably the most common and annoying hazard with which security and police personnel will be concerned. It can become such a financial menace and detriment to import-export operations that a large portion of police-security personnel efforts will have to be devoted to its control. Pilferage, particularly petty pilferage, is frequently difficult to detect, hard to prove, and dangerous to ignore.

The words *pilfer, pilferer,* and *pilferage* are used throughout this book in the sense in which they have come to be accepted by the industry rather than in the dictionary sense. Thus, they include the meanings of *steal, thief, theft, larceny,* and such terms, and apply not only to petty theft, but to theft of any quantity or monetary value.

To determine the severity of this hazard, there is need to determine the amount of loss that may be occurring. Unfortunately, this is not always an easy task. Accounting methods may not be designated to pinpoint thefts; consequently, such losses remain undisclosed or are lumped together with other shrinkage, thus effectively camouflaging them for all practical purposes.

One of the most common inventory methods is to conduct periodic inventories of property and to assume that unaccounted-for inventory loss is due to theft. This is a convenient but deceptive and dangerous device because theft is only one of many causes for inventory shrinkage. Failure to detect shortages in incoming shipments, improper stock usage, poor stock accounting, poor warehousing, improper handling and recording of defective and damaged stock, and inaccurate inventories will cause inventory losses that may be inaccurately labeled as pilferage. No physical security program, however efficient, will completely eliminate losses from these causes.

In some cases, inventory losses may be impossible to detect because of the nature and quantities involved. Stock inventory records may not be locally maintained, or there may be no method for spot-checks or running inventories for discovering shortages as they may occur. This is an undesirable situation and should be corrected where possible. In any case, the degree of risk involved can be determined only by analysis of the relative vulnerability of each area or activity of a terminal to the hazard of pilferage. In order to do this, it is necessary to consider the problem of who is likely to steal and what these people are most likely to take.

Profile

There are two types of pilferers who police, management, and security personnel must be prepared to counteract, or at least recognize, so that proper physical security measures may be taken against them. These are casual pilferers and systematic pilferers.

A *casual pilferer* is one who steals primarily because he is unable to resist the temptation of an unexpected opportunity and has little fear of detection. There is usually little to no planning or premeditation involved in casual pilferage, and the thief normally acts alone. He may take items for which he has no immediate need or foreseeable use, or he may take small quantities of supplies for use of family or friends, or for use around his house. The degree of risk involved in casual pilferage is normally slight unless very large numbers of persons are involved. Casual pilferage will occur whenever the individual feels the need or desire for a certain article and the opportunity to take it is provided by poor security measures. Though it involves unsystematic theft of small articles, casual pilferage is nevertheless very serious and it may have a great cumulative effect if permitted to become widespread, especially if the stolen items have a high cash or potential high cash value. There is always the possibility that casual pilferers, encouraged by successful theft, may turn to systematic pilferage. Casual pilferers are normally employees and usually are the most difficult to detect and apprehend.

A *systematic pilferer* is one who steals according to preconceived plans, and steals any and all types of supplies for the purpose of selling them for cash or bartering them for other valuable or desirable commodities. He may work with another person or with a well-organized group of people, some of whom may even be in an advantageous position to locate or administratively control desired items, or remove them from storage areas or transit facilities. The act of pilferage may be a one-time occurrence, or such acts may extend over a period of months or even years. Large quantities of supplies with great value may be lost to groups of persons engaged in elaborately planned and carefully executed systematic pilferage activities. Systematic pilferers may or may not be employees of a tenant. If they are not, they frequently operate in conspiracy with such employees.

Pilferage may occur anywhere. Even supplies that are stationary in permanent or semi-permanent storage areas or warehouses are vulnerable to theft if adequate precautionary measures are not taken—and vulnerability increases as supplies become more mobile. New and greater opportunities for pilferage are present when supplies are being transported in trucks or ships. The greatest vulnerability and the widest variety of opportunities occur at the various points where supplies are transferred from one means of transportation to another, or from storage to transportation and vice versa. It should be noted that anyone may be a pilferer. Where *need* or *desire* exists, and *opportunity* is presented, theft is almost sure to result.

Pilferage Targets

Both the casual and systematic pilferer have certain problems to overcome in order to accomplish pilferage objectives. The pilferer's first requirement is to locate the item or items to be stolen. For the casual pilferer this may be accomplished through individual search or even accidental discovery. In systematic pilferage, more extensive means are generally employed. These means may consist of surveillance by members of the group, or the checking of storage or documents by those who have access to them.

The second requirement confronting the potential pilferer is the manner in which he can gain access to the desired items and gain possession of them. This may involve something as simple as breaking open a box or it may be as complex as surveying such security factors as physical safeguards or guard procedures for weaknesses, attempting to bribe guards, altering or forging shipping documents or passes, or *creating disturbances to divert the attention of police while the actual theft is taking place.*

The third requirement involves the problem of removing the stolen items to a place where the thief may benefit from his act. Articles of clothing may be worn to effect this; small items may be concealed in any one of many possible places on the body of the thief or in vehicles; and, through falsification of documents, whole truckloads of cargo may be removed from their proper place without immediate discovery.

Finally, to derive any benefit from his act, the pilferer must use the item himself or dispose of it in some way. The casual pilferage of supplies is intended primarily to satisfy the need or desires of the individual thief. The systematic pilferer will usually attempt to sell the material, probably through fences, pawnbrokers, or other operators. The detection of use or disposal offers a chance to prevent further similar pilferage through investigation and discovery of the means used to accomplish the original theft. Similarly, each of the problems faced by would-be pilferers offers opportunities for constructive preventive measures. Careful study of the possible opportunities for the pilferer to solve his problems is essential in security work.

The primary concern of a systematic pilferer in selecting a target is its monetary value. Because he steals for personal profit, the systematic pilferer will look for items from which he can realize the greatest financial gain. This means that he must have or be able to find a ready market for items of property that he may be able to steal. He pilfers small items of relatively high value such as coffee, valuable metals, or radio and television sets. However, we cannot discount the possibility that a systematic pilferer may, if the profit is substantial, select a target which is of great size and weight. As a general rule, bulk storage areas contain most of the material that may be selected by systematic pilferers.

The casual pilferer is likely to take any item that is easily accessible. Because he normally will remove the item from the area by concealing it on his person or in his privately owned automobile, size is also an important consideration. Monetary value and available markets are not of any great concern to the casual pilferer because he usually does not have any idea of selling the property he steals. He normally uses the item for his own personal use. Any property not secured or under surveillance, or small enough to be hidden on the person or otherwise removed from the area by commonly available means is subject to casual pilferage. Storage areas containing loose items are more likely to tempt casual pilferers than are bulk storage areas.

Methods of Pilferage

There are many ways by which pilfered items may be removed from warehouses and sheds. Because the motives and targets likely to be selected by systematic and casual pilferers are very different, the methods of operation for each will likewise be dissimilar. As stated above, the casual pilferer will steal whatever is available to him and generally will remove it from the warehouse or shed by concealing it on his person or in his automobile. The methods of the systematic pilferer are much more varied and complex. The means that he may employ are limited only by his ingenuity. The following are cited as examples.

Shipping and Receiving Operations. These facilities are extremely vulnerable to systematic pilferage. It is here that warehouse, shed, and dock personnel and truck drivers have direct contact with each other and readily available means of conveyance, which offers a tempting opportunity for collusion. Although most truck drivers are honest, many of them do succumb to temptations. The driver may be operating in collusion with a receiving clerk who will certify the receipt of property which was disposed of prior to the driver arriving at the facility. A warehouse employee may provide a truck driver with property and assist in concealing it aboard the truck for unauthorized removal from the warehouse or shed. Employees may assist a truck driver in removing property by executing a fictitious invoice which appears to be legitimate when inspected by security personnel. The driver's responsibility starts when the truck is loaded and the cargo is receipted for; it ceases when the truck is unloaded at destination and the receipt is signed. The distribution and security of the load on the truck are also the responsibility of the driver and the agency that has custody of the material.

Railways. Employees assigned to switching duties on the facility can operate in a similar manner, but with more difficulty because a railway car normally cannot be directed to a desired location so that stolen property can be removed. Additional confederates will usually be required to transfer the stolen goods from the railway car, at some point or siding outside the facility, into some other means of transportation for removal to the ultimate destination. This increase in the number of persons involved will reduce profits and increase the chances for discovery and apprehension. The carrier's responsibility commences when the loaded and sealed railroad car is coupled to the engine. The carrier's responsibility ceases when the railroad car is spotted in the consignee's yard or unloaded at its destination.

Trash Disposal and Salvage Activities. These activities offer excellent opportunities to the systematic pilferer to gain access to valuable material. Property may be hidden in waste material to be recovered by a confederate who has the duty to remove trash from the facility. Serviceable or even new items of equipment or material may be classified as salvage by dishonest employees who are operating in collusion with other persons working in or having access to salvage disposal.

Other Methods. Systematic pilferers may remove property from tenant areas by such methods as throwing items over fences to be retrieved at a later time; packaging property and sending it to outside addresses through mail channels; collusion with security personnel; loose-fitting clothing that can be worn to conceal small items; and removal on vehicles belonging to outside contractors and vendors.

Measures for Control of Casual Pilferage

Specific measures for prevention of pilferage should be based on careful analysis of conditions at each vulnerable area. The most practical and effective method for controlling casual pilferage is the establishment of psychological deterrents. This may be accomplished in a number of ways.

One of the most common means of discouraging casual pilferage is to bring to the attention of all employees the fact that they may be apprehended if they attempt the illegal removal of property. Care must be taken to ensure that personnel are not demoralized or their legal rights violated by oppressive physical controls or unethical security practices.

An aggressive management security-education program may be an effective means of convincing employees that they have much more to lose than they do to gain by engaging in acts of theft. Case histories should be cited where employees were discharged or prosecuted for pilferage. It is important for all employees to realize that pilferage is morally wrong no matter how insignificant the value of the item taken.

It is particularly important for supervisory personnel in warehouses and sheds to set a proper example and maintain a desirable moral climate for all employees. All employees must be impressed with the fact that they have a responsibility to report any loss to proper authorities.

Adequate inventory and control measures should be instituted to account for all material, supplies, and equipment. Poor accountability, if its existence is commonly known, provides one of the greatest sources of temptation to the casual pilferer. The identification of all tools and equipment by some mark or code, where feasible, is necessary so that tenants' property can be identifiable.

Measures for Control of Systematic Pilferage

Unlike the casual pilferer, the systematic thief will not be discouraged by psychological controls. Nothing short of active physical security measures will be effective in eliminating losses from this source. Some of these would be to:

1. Establish an effective guard surveillance of all entrances and exits in cargo areas.
2. Establish an effective package and material control system.
3. Locate parking areas for private vehicles outside the perimeter fencing of the cargo work area.
4. Eliminate potential thieves during the hiring procedure by careful screening and observation.

5. Investigate all losses quickly and efficiently.
6. Establish an effective key-control system.
7. Establish *adequate* police-security patrols to check buildings, grounds, stored pallets, and locations that might be used for the clandestine storage of property that has been removed from its proper location.
8. Install mechanical and electrical intrusion-detection services where applicable and practical.
9. Establish appropriate perimeter fencing, lighting, and parking facilities and effective pedestrian, railway, and vehicle gate security controls.
10. Store bulk quantities of highly pilferable stock in enclosed security areas and distribute from there to the consignee.
11. Establish accurate methods of taking physical inventories and of accounting for cargo received, delivered, and salvaged.
12. Establish close liaison with other law-enforcement and intelligence services where possible.

Prevention

Prevention of pilferage requires alertness and aggressiveness on the part of all police and security personnel, careful planning and implementation of physical security measures, and an extensive security education program for *all employees* of air and marine terminals. The prevention of pilferage is a continuing process that must be given careful attention. All means available, from the use of informants to the installation of security devices, should be employed to ensure timely information of illegal activities that may lead to serious financial losses.

PHYSICAL SECURITY

Approximately 80 percent of thefts will occur in the physical plant (building and warehouse) itself. With this in mind, all cargo-handling and storage facilities should provide a physical barrier against unauthorized access to its contents. The organization failing to build security into its physical plant proffers an invitation to theft. To offset this invitation, a simple three-step approach to physical security is essential.

1. Construct defense against theft.
2. Determine means to penetrate those defenses.
3. Make necessary modifications to improve the defenses.

Buildings

All buildings should be constructed or renovated to specifications and in such a manner as to resist unlawful entry. This integrity must be maintained by means of periodic inspection, repair, and upgrading. In order to maintain inaccessibility to physical plants, the following procedures are suggested.

1. All exterior doors and windows should be equipped with locks or locking devices that have been properly installed.
2. Windows accessible within 18 feet of ground level should be safeguarded by means of safety glass, wire mesh, or bars.
3. All glassed-in areas where shipping documents are processed should be treated in a similar manner as above.
4. All delivery and receiving doors should be constructed of steel or similar material to prevent or deter illegal access. When not in use, all doors shall be closed and locked.
5. If conventional means of security are impractical or security personnel insufficient, detection or alarm systems should be incorporated. Thorough inspections should be conducted to ensure that no means of entry can be made through floors, roofs, or adjacent buildings undetected.
6. Only authorized personnel should be allowed access to the physical plant, high-value storage areas, and paperwork areas.

CARGO LOSS PREVENTION EQUATION

Motive + Opportunity + Means = Crime (Cargo Loss)
Motive. Cannot be controlled.
Means. Difficult to control.

Opportunity. Security personnel must attack this portion of the formula and reduce opportunity as much as possible to effect the end result.

Any loss-reduction program must discourage, observe—identify, and respond—control.

Discourage. By means of:

1. Fencing/Razor Wire
2. Signing
3. Lights
4. Natural Barriers
5. Patrol
6. Control Points
7. Security Devices
8. CCTV

Observe-Identify. By means of:

1. Personal Recognition/I. D. Badges
2. Employee Identification
3. Vehicle Identification
4. Surveillance
5. Company Uniforms
6. Patrols

Respond-Control. All the discouragement and identifying is useless unless you *challenge* the ones who appear in an area that is restricted.

Fencing

Proper fencing is an integral part of security. It is utilized to enclose areas surrounding cargo storage structures, support buildings and exterior cargo storage areas. It should be situated so as to provide ample maneuvering space for pickup and delivery vehicles; a clearance of at least 20 feet should be maintained to prevent the use of vehicles or cargo to surmount the fence. Regular inspection of fencing shall be conducted to ensure its integrity and to allow prompt repair of damage. The benefits of proper fencing include the following:

1. Deters accidental entry to an area.
2. Deters unauthorized entry for criminal purposes.
3. Delays entry into areas by criminals determined to commit a crime, making apprehension likely.
4. Restricts pedestrian and vehicular traffic to designated entrances.
5. Curtails pilferage by employees by creating a psychological fear of apprehension.

In order to provide uniformity in the construction of fencing, the following are recommended specifications:

1. Install chain-link fencing, with mesh openings not larger than 2 inches and made of 9-gauge (or heavier) wire with twisted and barbed selvage top and bottom. The fence shall be at least 8 feet high. If the fence is constructed on uneven ground, the height will be adjusted to provide an effective 8-foot fence at all points.
2. The fence shall be topped with a 2-foot barbed wire extension consisting of 3 strands of barbed wire or razor ribbon, properly spaced at a 45-degree angle to the vertical.
3. Fence posts should be placed inside the fence and secured in a cement foundation at a minimum of 2 feet.
4. Ensure that persons or objects cannot pass beneath the fencing by providing one of the following barriers:
    a. Cement aprons not less than 6 inches thick.
    b. Frame piping.
    c. U-shaped stakes driven approximately 2 feet into the ground.
5. Prevent compromising the fence line. Prohibit the placing of any commodity (containers, dunnage, cargo, vehicles, etc.) in a position that will allow a bracing of the fence line.
6. When warranted, install anchored concrete bumpers or fence guards to prevent damage from vehicles, and create a corridor providing a clear view of the fence line to management and security personnel.

Gates

The number of gates and access points should be limited to the minimum number of required for entrance and exit. The construction of the gates should be of the same quality as the fencing. All gates utilized by vehicles or personnel should be manned or under observation at all times.

Recommended specifications in this area are:

1. All gates should be equipped with deadlocking bolt or an equivalent device that does not require use of a chain. Hardware connecting the lock to the gate should be of the type to withstand constant use and any attempts to defeat it (case hardened).
2. Cantilever-type gates should be constructed that can be secured with minimum ground clearance when closed.
3. Separate gates should be constructed for the use of personnel and vehicular traffic. Personnel should be required to pass a screening point, if only for visual inspection.

Gatehouses

Gatehouses at all entrances and exits of facilities handling a substantial volume of high-value cargo should be manned during business hours. These gatehouses should be situated in the centers of the ingress and egress roadways.

Areas of concern for the placement and equipping of gatehouses include the following:

1. Gatehouses should be situated in from the gate so that exiting vehicles may be halted and examined on terminal property.
2. All gatehouses should be equipped with telephones or other communication devices and alarm buttons.
3. The area surrounding gatehouses should be cleared of any items that restrict the guard's line of vision.
4. Signs advising drivers and visitors to stop and of the conditions of entry onto agency property should be posted prominently on the exteriors of all gatehouses. Included in these conditions should be a notice that all vehicles and personnel entering are subject to search. All signs should be international in design.

Parking

Private passenger vehicles should be prohibited from entering or parking in cargo areas and immediately adjacent to cargo storage buildings. Control of access to employee parking areas should be supervised.

Further areas of consideration:

1. Parking areas should be situated outside of fenced operational areas, or, if impractical, a substantial distance from cargo-handling and storage areas or buildings. All parking areas should be lined and signed.
2. Employees exiting to the parking area from a cargo facility should be required to pass through a controlled area under the supervision of management or security personnel.
3. Discourage employees from visiting their autos at odd times. If the trip is necessary, require notifications to be made to management or security personnel.
4. Parking in employee parking areas should be restricted and by permit only. The permit may consist of a numbered decal, tag, sign, etc., placed in a uniform location on the vehicle. A record will be maintained of all permits issued consisting of: permit number; vehicle registration number; model, make, and year of vehicle; and person of issuance.
5. Temporary permits, to allow parking in designated controlled areas, will be issued to vendors and visitors.

Lighting

There should be controlled adequate lighting for all areas. Agency facilities should be clearly illuminated. The following are recommended criteria.

1. Entrances, exits, and gatehouses should be sufficiently lighted to assure proper visibility of persons and vehicles approaching the area. Minimum illumination: 2.0 foot-candles, controlled by photo cell or timing device.
2. Perimeter lighted areas should be clearly illuminated so as to permit easy detection of an intruder. Minimum illumination: 1.0 foot-candles.

3. Electric lines serving lighting on the perimeter fence should be underground or at a height not easily accessible. Light fixtures should be accessible only from within the fencing, and covered with a protective screen or device.
4. Light controls should be activated by an automatic timing device. This will reduce electricity consumption and minimize bulb changes.

Loading Docks

Many acts of theft and pilferage occur between the dock and the terminal warehouse. To lessen or eliminate these occurrences, the recommendations are:

1. Separate receiving and shipping records.
2. Maintain restrooms, telephones, etc. nearby, but separate from the dock and/or cargo offices.
3. Have supervisory personnel present during lunch and coffee breaks if these are taken in the work area.
4. A security cage shall be maintained for high-value cargo.
5. Supervisory offices should be constructed at a height to allow an overview of the dock.

Buffer Zones

Cleared areas should be maintained on both sides of a fenced area to provide an unobstructed view of the fence and the ground adjacent to it. Police vehicles should patrol as close as the terrain allows.

1. A buffer zone of 20 feet or greater should be established between any fence or exterior buildings and parking areas. This is essential in areas where large numbers of foreign cars or other vehicles are stored.
2. In areas adjacent to warehouses and sheds, a buffer zone of 10 feet should be created around the entire building. This should be enforced *at all times*.
3. Under no circumstances should any vehicle be permitted to park against a building at any time, unless loading or unloading cargo at a platform.

4. No vehicle, loaded or empty, should be left parked against a loading platform overnight.
5. When it is impossible to have adequate buffer zones because of property lines or other features, additional fencing may be necessary, as well as additional lighting and more frequent police patrols.
6. If properly installed and maintained, buffer zones should afford the police patrols a clear view along fenced areas. They will also permit the officer to observe all doors, windows, and other entrances to a building, day or night.

Debris, Stored Pallets, and Abandoned Vehicles

Debris piled against warehouses and sheds affords natural cover for hidden stolen property. Pallets stored against warehouses and sheds provide pilferers with a place to cache stolen cargo. Placed near doors and windows, they encourage pilferage. Abandoned vehicles, unused and parked in close proximity to buildings, afford cover for stolen cargo and material. Containers offer natural cover for illegal activity and appear, in many cases, to be strategically placed to obstruct the view of building entrances to police and security guards.

1. Enforce strict control over tenants' use of the area outside his warehouse, shed, fences, and operating area.
2. Immediately remove broken pallets from loading and receiving areas. All broken pallets should be stored outside the buffer zone at least 10 feet from a warehouse and fences until disposed of.
3. New pallets should be stored inside a building or outside neatly stacked at least 10 feet from the structure.
4. Enforce removal of abandoned trucks, broken containers, and disabled machinery.
5. No machinery, pallets, containers, or equipment of any description should be stored within a buffer zone.
6. Institute a general cleanup campaign and apprise the operators of the reasons for this campaign.
7. Eliminate as many pallets and containers from outside warehouses as you possibly can. This would lessen the chances of

pilferage, fire, and sabotage. All areas should be kept free of dirt, waste material, and debris.

Guarded Parking Areas

Physical protection recommendations for guarded parking areas include good lighting, antitheft fencing, buffer zones, and a general cleanup.

Once a security parking area is constructed, these basic steps can be followed to deter theft from this area.

1. Armed guards and alarm systems should be used.
2. Trucks should be parked in the lot back to back.
3. Use fifth-wheel (king-pin) locks for maximum security.
4. Check seals frequently to ensure that they remain intact.
5. Provide an alarm system or telephone between the guard force and the police.
6. The police or management should make a cursory check to ensure the presence and safety of guards.
7. Inspect all truck drivers' credentials and pickup or delivery orders before releasing loaded trailers. This includes a routine check of drivers' licenses to ensure the identity of the drivers.
8. Investigate the possibility of taking "Regiscope" photographs of drivers, papers, etc.
9. If union problems are encountered, it should be stressed that this is done for the protection of all the legitimate truck drivers who suffer when a hijack occurs.

Procedures for Gateman

1. Trucks must be checked for the proper name displayed on the side in stencil, paint, or decal. Crayon, magic marker, masking tape, and other nonpermanent placards are not acceptable as properly identifying trucks when presented at the gate.
2. Temporary placards, as used by lease operators or owner operators, may be accepted if the operator of the truck can produce an acceptable *trip lease agreement* showing affiliation with the truckman of record.

3. All leased-equipment drivers must produce lease papers and driver's license prior to being passed through the gate.
4. Straight trucks, where the body and cab are one unit, may be admitted when the name of the company appears on the body of the truck, providing that it is the company of record. All others must have a lease agreement or letterhead authorization to be approved.
5. Truck registration must be presented at the gate and checked by gatemen for matching vehicle plates.
6. Examination of *all vehicles* is necessary upon exiting. This is especially true when trucks have entered with shipping cargo and exit with a partial load.
7. Any truckers who do not comply with any of the procedures should be refused entry. If, after a truckman gets a gate pass, a discrepancy is observed, the pass shall be voided.

Checkpoints

Many people have expressed interest in establishing checkpoints at given locations where physical examinations of trucks and cargo can be performed, thereby uncovering stolen items and curtailing future attempts to steal large quantities of cargo. This can only be accomplished within the frame of local laws. Therefore, individual research is necessary.

Employee and Visitor Parking and a Buffer Zone

In conducting research interviews, we observed laxity on the part of many operators in controlling vehicular parking within or near their complexes. Needless to say, vehicles parked immediately adjacent to a work area are ideal cache points for the systematic or casual pilferer. The need for a parking area away from the building at an inactive site is stressed to management. At some locations it is a common practice to park loaded or unloaded trailers against the side of the building, leaving no space between the rear of the vehicle and the warehouse.

Criminals desiring to gain entry into a warehouse, shed, or other building can cut through the siding of the building and back a trailer against the aperture, the trailer offering a cover for illegal activity. One operator, upon arriving at his place of business on a Monday morning, found the side of his warehouse cut open and a loss that he estimated required three full trailers to remove. The modus operandi was an eight-foot hole cut into the side of the building.

Protective Lighting

Protective lighting is considered a necessity, and when properly employed, it should reduce the advantages of concealment and surprise a determined intruder. Each area of a facility presents its own particular problem based on the physical layout, terrain, and protective requirements.

Protective lighting should enable a police or guard officer to observe activity around or inside a given complex without disclosing his own presence. Two basic systems, or a combination of both, may be used to provide practical and effective lighting. The first method is to light boundaries and approaches; the second is to light the area and structure within the general boundaries of the property.

To be effective, protective lighting should:

1. Discourage or deter attempts at entry by intruders. Proper illumination may lead a potential intruder to believe detection is inevitable.
2. Make detection more likely if entry is attempted.

BUILDING SECURITY

Truck drivers should not be permitted to congregate in the general offices of any terminal other than as required to conduct their business. They should be restricted to loading platforms, in close proximity to their trucks, to protect their loads.

The dispatcher's office is the center of operations, and admittance should be restricted to the terminal manager, assistant terminal manager, dispatcher, foreman, and clerk in pursuit of company business.

Only authorized personnel should be permitted in the warehouse. It is the foreman's and/or the company supervisor's duty to challenge all others by determining the nature of business and reason for their being there. The warehouse should be well lighted for the purpose of observing functions of employees and spotting unauthorized persons in the area. Warehouse bay doors and their locking devices should be regularly examined for defects and immediately repaired. Overhead doors should be lowered and locked when unattended (including coffee and meal breaks). During warm season or in areas where the temperature remains consistently high, expansion, scissors-type doors or gates can be used to allow air circulation for employee comfort.

Key control must be maintained and responsibility identified. All locks on the terminal should be opened by three master sets of keys. Master sets should be in the custody of the terminal manager, the security director, and the roundsman. In the event of the loss of a master key set, all locks associated with the lost set must be rekeyed in order not to compromise security.

There should be no individual keys in the possession of terminal personnel. The set in the possession of the roundsman must be kept locked in the security office. All terminal offices should be secured at the end of the day. Security personnel will admit cleaning personnel and secure doors on completion of cleaning. They will open all doors in the morning.

The crib area in the warehouse to be used for high-value cargo will vary in size depending on the dimensions, volume, and quantity of cargo stored therein. It should be constructed of heavy-gauge no. 9-wire cyclone fencing, fixed by tack welding to the steel support posts. The crib of an average warehouse should measure approximately 80 feet in length, by 40 feet in width, and should be a minimum of 18 feet high or fitted to the ceiling. In the event that the ceiling is greater than 18 feet, an additional 2 feet of barbed wire (3 strands) set on an angle should be added.

In lieu of the barbed wire "topping" the cage or crib may be totally enclosed with equal wire fencing on sides and top. A 15-foot sliding-type gate, with only enough clearance at the bottom to allow easy sliding, should be secured by appropriate locking hardware and padlock. Sections of concrete curbing, such as concrete parking-lot bumpers, should be secured to the warehouse or shed floor, along the inner perimeter of the crib fencing.

In all cases a vault or large safe should be available for storage of high-value cargo. Alarm systems, both silent and audible, are recommended as primary unauthorized-entry alert systems. Closed-circuit television installations are recommended as well as a guard service to monitor the television pictures and to maintain physical security inside the high-value room or crib. A high-value room or crib logbook should be maintained, listing all personnel who enter the room and the items either left for storage or removed.

SECURED TRUCK STORAGE AREA

A basic requirement of a safe storage area for loaded trucks is that it be fenced and well lighted. It must be kept in mind that physical security recommendations will not prevent theft caused by collusion between the tenant and the trucker, or among their personnel.

Fencing of Security Storage Area

Observation and research indicate that the fence should be at least 8 feet high and topped with barbed wire. The fence should enclose the entire secured parking area. It is recommended that the fence be mounted on a *concrete wall* or *base* to preclude the possibility of a loaded truck being driven through it. This base should be high enough to stop a tractor and trailer loaded with cargo.

1. Windows overlooking the security area should be covered with a heavy wire screen to deter unauthorized entry. Such screening will also prevent cargo from being passed into the building from the security lot by drivers, security personnel, and other dishonest employees working with a confederate inside.

2. In no case should a car or other vehicle be allowed to park next to a fence. It is recommended that two barriers be erected, one inside the lot and the second outside to form a "safe zone" or buffer zone of at least 10 feet. This can be accomplished by the use of performed concrete slabs fastened to the paving material of the lot. The purpose of the buffer zone is to deter pilferers from gaining entrance to the lot from atop a vehicle parked next to the fence. It also prevents cargo from being thrown over the fence to a waiting confederate who would place it into a vehicle parked nearby. Most important, it affords patrolling police and guard personnel an unobstructed view of the security lot; this view substantially reduces the chance of a successful hijack. If unauthorized entry is to be deterred, the buffer zone is a must. Otherwise, the effectiveness of the fence is substantially reduced.

3. Vehicles and pedestrians should be restricted to designated gates. At the entrance gates an *effective security officer force* is needed to control traffic into and out of the lot. At night and on holidays gates should be kept closed and locked.

4. *At no time* should debris or pallets be allowed to accumulate near a fenced area. It offers natural cover where stolen cargo can be cached for later removal by the thief or a confederate.

Lighting

Lighting units for restricted-area perimeter fence lighting should be located a sufficient distance within the protected area, and above the fence, so that the light pattern on the ground will include an area both inside and outside the fence. Generally the light band should illuminate the restricted-area barrier and extend as deeply as possible into the approach area. Factors such as adjacent waterways, highways, and railroads may limit the depth of the light band.

1. Appropriate protective lighting should be employed at all pedestrian and vehicular entrances where security officers or other personnel are stationed to check identification of personnel and bills of lading or other papers prior to release of valuable cargo.

2. Periodic inspections should be made of all electrical circuits to replace worn parts. Lights and lamps should be kept clean and properly aimed.

3. It is recommended that a lighting expert be consulted to determine the type of lighting system to be used.

Vehicle Storage Within the Compound

1. Keep all vehicles locked when unoccupied. Require drivers to shut down equipment, especially when diesels are involved.

2. *Lock and seal* all trailers stored in the security lot.

3. Install steering wheel locks on tractors.

4. Install fifth-wheel kingpin locks on trailers.

5. Park trailers back-to-back in designated areas.

6. Install audible alarm devices on "hot" loads (cigarettes, cigars, whiskey, appliances, etc.). Portable alarm devices are available that sound an audible signal should someone gain entrance to the security lot and attempt to tamper with a loaded trailer.

Release of Valuable Cargo

The security officer should demand the identification of the driver and necessary papers prior to releasing the loaded trailer and allowing the driver to proceed. If the *slightest* doubt exists in the security officer's mind or he becomes suspicious, the driver should not be allowed to remove the load until cleared by a higher authority. If a driver attempts to flee upon being questioned, the guard should note data such as tractor number, registration of vehicle, and name, if any. This information should be relayed to the police as soon as possible for follow-up by detectives. Naturally, security officers assigned to a cargo storage lot must be alert and exhibit enthusiasm in the performance of their duty if a security storage area is to succeed.

1. Require drivers to show a driver's pickup order, as well as equipment load and serial numbers, prior to leaving the security lot. This should reduce the number of thefts by a company's own driver-employees.
2. Install a device that photographs the driver and his papers before he leaves with the load.

Thefts by Company Employees

Well-lighted security storage areas may stop a sneak thief, but won't deter a company employee from spotting a trailer and supplying an accomplice with bogus papers. To deter this, we offer a few suggestions:

1. Make background checks of all employees by investigating their references and spot-checking employee habits. (Is he in debt? Does he gamble? etc.)
2. *Check the criminal history of* all employees working as security officers.
3. Supervise closely all new employees after the probationary period. (A truck thief is unlikely to make his move during his period of probation.)

4. Immediately discharge all employees convicted of theft of cargo, vehicles, or other property. This policy should be made unmistakably clear to a prospective employee before he is hired.
5. Tighten office controls on the storage and distribution of vital materials such as invoices, manifests, and shipping orders. In many cases, bogus bills of lading have allowed a driver to clear a police roadblock with stolen cargo.
6. The tenant or operator must impress a new employee with the idea that all employees will be prosecuted if they are caught pilfering. Many employees at a cargo facility feel that stealing from their employer is a "fringe benefit." If the employer makes a complaint and prosecutes, his pilferage will decline.

Open Storage of Cargo

Open storage is normally used only for those items of cargo not subject to damage by weather conditions, and bulky, nonperishable items that are not sensitive in the sense of being easily pilfered or disposable (lumber, mica, salt, etc.). However, open storage may include domestic and foreign cars, which are vulnerable to pilferage and vandalism.

1. When property is stored in open areas, it should be properly stacked and placed within, away from, and parallel to fences in order to permit an unobstructed view of the fences by police and guards. Stacks and lines of cargo should be a minimum of five feet from the fence. Aisles between stacks or lines should be wide and straight.
2. Fixed-position lights in a storage area should be of a diffused type to eliminate deep shadows.

Covered Storage

"Hot" cargo should be kept separate from other material. Items such as food, meat, cigarettes, cigars, whiskey, precious metals, radios, etc., fall into this category. The best solution would be to store such cargo in a separate building. In cases where this is not possible, a cage or crib may be constructed within a warehouse. The floors and roof of such an enclosure must be comparable in strength to its walls. In one valuable-cargo crib it was discovered that solder wire had been substituted for the heavy-gauge wire normally used to secure a cyclone (fence) enclosure. It was found that the solder wire could be removed with the fingers, access gained, cargo removed, and the solder wire replaced in minutes.

1. The same principles of even stacking and aisle space recommended for use in open storage are applicable to covered storage. Lighting should be used to reduce deeply shadowed areas to a minimum.
2. In warehouses where stored items are moved frequently, more emphasis should be placed on security guards than on structural protection, to prevent pilferage by workers.

Protection of High-Value Loads

An area capable of being locked, sealed, or otherwise secured for storage of high-value cargo must be provided at all cargo-handling buildings. When such cargo must be transported a substantial distance to the security area, vehicles capable of being adequately secured must be used. The following areas of consideration are recommended.

1. Construct special security areas (rooms, cribs, or vaults) so as to resist forcible entry from all sides and from overhead and underneath. When fencing issued, it shall be a minimum of 18 feet in height, constructed of no. 9 gauge wire, tied with heavy-gauge binders, and closed at the top.
2. Such cargo should be predestined. Immediately upon receipt it should be taken to the security area, followed by a checker who will supervise the transfer of the cargo to the security area, and the movement of such cargo should be conducted in such a manner as to guarantee a controlled flow of materials.
3. Security areas, where possible, should be located to assure constant observation by management and/or security personnel. If this is not possible, installation of alarm systems or inspections and inventories at frequent, unscheduled intervals, are recommended.
4. Merchandise should be released only in the presence of authorized supervisors and/or security personnel.
5. All movements of merchandise (entering or exiting the security area) should be logged. Such a log should contain: date, time,

condition of cargo upon receipt by truckman, number of pieces, company making the pickup or delivery, and registration number of equipment used.

Handling and Storage

This section encompasses the physical handling (excluding sealing, which was reported previously) of varied systems. Although some of the information may seem repetitive, it was deemed necessary to enumerate in each instance.

Containerized Shipments

1. Where possible, secure containers by butting or "marrying" their door ends against each other. In no circumstances should butting be accomplished against perimeter fences or building walls if it compromises the protection provided by the fence or wall.
2. When containers are stacked, those containing high-value merchandise should be placed on top.
3. Mounted containers should be situated in a secure holding area where they may be observed by management and/or security personnel.
4. When containers are frame-mounted, the fifth wheel should be secured by a pin-lock that meets the minimum standards for locks and that is constructed to withstand normal abuse from equipment. Management and/or security personnel shall be held responsible for the storage and control of pin-locks.
5. Movement of containers entering or exiting holding areas should be logged containing date, time, name of truckman and company expediting the movement, and registration number of equipment utilized.
6. Yard hustlers for each company should be painted a distinctive color for ready identification outside of their normal areas.

Vans, Trailers, and Trucks

1. An accurate and recorded tally of the shipment should be made.
2. A shipment should be delivered to the consignee as promptly as possible.

3. Delivery receipts should contain the complete legal signature of the consignee and the time and date.
4. Check all empty vans outbound.
5. Spot-check tractors, vendors, and service vehicles as traffic permits.
6. Account for the amount of time the vehicle is within terminal property.
7. Paperwork in the driver's possession must correspond with data from the vehicle and the shipment.

Rail Cars

1. A checker shall be present at all times.
2. Upon completion of unloading, the car should be checked to assure that it is empty.
3. When tiered for storage, the cargo should be rechecked.
4. The supervisor present should countersign all tallies.
5. Loaded rail cars should be sealed as recommended.

GATE PASS PROCEDURES: LESS-THAN-CONTAINER LOAD (LCL)

Receiving Procedures (Delivery by Trucker Into the Port)

1. The driver should park his truck in the LCL truck-holding area and should walk to the gatehouse, where he should present his papers, delivery order, and vehicle registration, and then he should be photographed.
2. The gateman should insert the following information on the pass, which is color coded (example: green for outgoing cargo, yellow for incoming cargo:
   a. Trucking company and plate (registration) number.
   b. Driver's name.

   He then should time-stamp the pass to indicate the time of arrival at the holding area. The gateman then should refer the driver to the proper delivery office. All information is then logged in the gate logbook.

3. The driver should walk with his gate pass and dock receipt to the receiving office, where he will be routed by the receiving clerk. The receiving clerk should inspect the paperwork to ascertain if everything is in order.
4. The driver then should walk to the dock boss' office on the unloading platform, and should present his gate pass and dock receipt. The driver at this point is told to go to the drivers' room and he will be notified as to when he will be unloaded. The dock boss will check the dock receipt and will assign a checker to the unloading of the driver's truck. The driver is given back his gate pass and instructed to drive his truck into the yard and back in to the assigned door. The dock receipt is given to the assigned checker along with a tally sheet.
5. The driver should return to his truck and prepare to enter the compound via the gate. At the gate, he should give the gateman his pass, which is inspected. If found to be correct, the pass is time-stamped to indicate time he entered the terminal, and then the driver is directed to the assigned door.
6. The driver should back his truck into the assigned door and should be met by the assigned checker.
7. The truck should be unloaded and a tally kept of cargo unloaded. Upon completion of unloading, the driver should pull away from the door, park his truck, secure it, and then return to the receiving office.
8. At the receiving office the driver should sign the pier papers and be given receipt along with the gate pass.
9. The driver, with his truck, then should depart the terminal and surrendering his gate pass to the gateman, who time-stamps the pass to indicate the time the driver and vehicle exited the terminal compound.
10. All trucks should be opened and inspected by the gateman upon leaving the terminal. If a truck has entered with a partial load, this should have been recorded, informing the gateman as to the correct number of pieces aboard the truck when it exits.

Delivery Procedures (Pickup by Trucker to Take Out of the Port)

1. The driver should park his truck in the LCL truck holding area and should walk to the gate house, where he should present his

papers, pickup order, and vehicle registration, and then he should be photographed.

2. The gatemen should insert the following information on the pass, which is color-coded (example: green for outgoing cargo, yellow for incoming cargo):

    a. Vessel.

    b. Trucking company and plate (registration) number.

    c. Driver's name.

    He then time-stamps the pass to indicate the time of arrival at the holding area. The gateman then should refer the driver to the proper pickup office. All information is then logged in the gate logbook.

3. The driver should walk with his gate pass and delivery order to the office, where he will be routed by the delivery clerk. The clerk should inspect the paperwork to ascertain if everything is in order. The driver is then directed to the U.S. Customs Office.

4. The driver at the U.S. Customs Office should present his pickup orders. Customs should check with its own files to see if all Customs rules and regulations have been complied with, and, if so, Customs will stamp the pickup order approved.

5. The driver then should walk to the dock boss' office on the loading platform and should present his gate pass and pickup order. The dock boss will assign a checker to the loading of the driver's truck. The driver is given back his gate pass and instructed to drive his truck into the yard and back into the assigned loading door. The pickup order is given to the assigned checker along with a tally sheet.

6. The driver should return to his truck and prepare to enter the compound via the gate. At the gate, he should give the gateman his pass, which is inspected. If found to be correct, the pass is stamped for the second time. The driver then should enter the terminal, where he is directed to the assigned door.

7. The driver should back his truck into the assigned door and should be met by the assigned checker. Prior to loading the truck, the checker will notify U.S. Customs on the platform and the cargo will be physically inspected. When it passes this inspection, the U.S. Customs officer will allow it to be loaded onto the truck.

8. The truck should then be loaded and a tally kept of cargo loaded. Upon completion of the loading the driver should pull away from the doors, park and secure his truck, then return to the delivery clerk's office.
9. At the delivery clerk's office the driver should sign the pier papers and should be given a receipt along with a gate pass.
10. The driver, with his truck, should then depart the terminal and surrender his gate pass to the gateman, who time-stamps the pass to indicate the time the vehicle exited the terminal area.
11. All trucks should be opened and inspected by the gateman upon leaving the terminal. If a truck had entered with a partial load, this should have been recorded, informing the gateman as to the correct number of pieces aboard the truck when it exits.

PROCEDURAL SECURITY

Personnel Screening

It is incumbent upon operators of cargo-handling facilities to process prospective employees through a screening procedure. This shall require all personnel who will have access to cargo areas to submit a detailed employment application that contains a photograph of the applicant and that lists his residences and prior employment for the preceding 10 years. If a lapse is evident, an affidavit should be obtained for the unaccountable time. (Was he is jail?) In addition, the following areas of consideration are recommended:

1. Require that all applicants be fingerprinted. This is a means of positive identification. It will deter applicants from filing under assumed names and concealing arrest records.
2. Verify all listed addresses.
3. Verify any prior employment. Be aware of any employment gaps.
4. Check all references.
5. Investigate the applicant's credit record and, if applicable, record of criminal connections.
6. Require those applying for "sensitive" positions to complete a bonding application.

7. Inform prospective, as well as current employees, of security regulations and the consequences attached to violations of those regulations.

Security Personnel

Security personnel may be employed by the operator of a terminal or contracted to a guard service. Operators of cargo-handling facilities should employ a security officer or assign an officer of the firm to assume the responsibilities for security. All operators handling substantial shipments of international cargo should provide guards to protect the cargo. To assure that services provided are sufficient, the following procedures are recommended:

1. In the absence of electronic surveillance systems, a sufficient number of security officers shall be employed to provide adequate security in relation to the size of the facility and the amount of cargo handled.
2. Train all security personnel, internal or contracted, in:
   a. Methods of patrol regarding terminals and warehouses.
   b. Identification of security problems and specific trouble areas.
   c. Report writing and record keeping.
   d. If required, the proper use of firearms or other equipment that may be furnished.
3. It is recommended that all guard forces be equipped with uniforms that are complete, distinctive, and authoritative in appearance.
4. Provide any equipment (firearms, vehicles, communications systems, etc.) that will be necessary for the successful performance of an effective security function.
5. Physical fitness should be a prime consideration in the selection of a guard force. Physical examinations should be required annually. Self-defense training similar to that of police agencies may also be required.
6. Each guard shall be supplied, by the agency or terminal operator, with a manual incorporating operating procedures, standards of conduct, and a definitive statement of what the management expects of him.

Communications

Adequate and reliable communications should be provided to enable contact among elements of the terminal security force and from the security force to local law enforcement. This can be provided in the following manner:

1. Security personnel should be provided with a telephone at fixed posts or two-way radio, intercom, or similar type of equipment that will provide voice communication within the terminal area.
2. Assured means (telephone, radio, special alarm line, etc.) should be arranged for summoning assistance from local police.

Employee-Identification System

All terminals handling a substantial volume of cargo should employ an identification-card system to identify personnel authorized to enter cargo and document-processing areas.

1. Information recorded on the identification card should include, at the very least:

    a. Employee's color photograph.

    b. Name and address.

    c. Social security number.

    d. Date of birth.

    e. Signature.

    The card should also display a reasonable expiration date, for example, after one year.

2. Each card should be assigned a control number, and should be laminated to avoid tampering.
3. Identification cards should be immediately recovered from employees at the time of termination of employment.
4. Each employee should be required to display his identification card in such a manner that it is visible at all times. This will facilitate and control access to the facility, cargo areas within the facility, and areas where shipping papers are processed.

SECURITY PERSONNEL

In the following section we list those qualities that we feel are necessary to an effective operation. Naturally, before selecting security officers, minimum physical and mental standards should be established.

Alertness. This quality determines the effectiveness of a security guard, and must be cultivated. A guard must watch for deviations from normal routine such as strange trucks or cars, or a person approaching from an area not normally used.

Judgment. Security instructions cannot cover each situation. The guard must learn to compare, discriminate, and formulate accurately, and to arrive at a sound decision.

Confidence. A guard must have confidence in himself, his fellow guards, and his superiors.

Physical fitness. A security officer works long hours when ships or planes are being unloaded and cargo removed. The guard should be of sound mind and body because the job, if done correctly, is demanding and he is exposed to the weather for long periods of time. Heavy drinkers or alcoholics have no place in the security field.

Appearance. A security officer should be properly uniformed at all times. At present the only uniform equipment some security officers own is a police-type cap. When completely and properly uniformed, the security officer is easily identified as a figure of authority. We feel that this alone would have a deterrent effect on pilferers.

Equipment. When given the responsibility of guarding valuable cargo, the security officer should be armed. People who are selected for this duty should exhibit fair proficiency with a weapon. A sneak thief will have second thoughts before attempting to challenge an armed security officer in good physical condition.

Tactfulness and self-control. A guard should be able to give instructions clearly, firmly, and concisely, but without arrogance. In addition, he must not lose control of his temper.

Background. Each security officer should undergo a record check by a responsible authority. This will ensure that valuable cargo will be entrusted only to persons of known responsibility and trustworthiness, and that collusion between thieves and guards will be held to a minimum.

THE SECURITY OFFICER

Training

We believe that security officer training objectives can be achieved with eight days of academic training and two days of on-the-job training with an experienced security officer.

Eight Day Security Officer Curriculum

Week One

First Day

0800–0900 Introduction to the Port and Employing Company

0900–1000 The Duties and Responsibilities of the Security Officer

1000–1100 Employee Rules and Regulations

1200–1300 Lunch

Fundamentals of Security and Fire Alarm Systems

1500–1600 Ethics and Courtesy

Second Day

0800–0900 Physical Security Systems I: fences, lighting, and locks

0900–1000 Other Law-Enforcement Agencies; FBI-guest FBI speaker

1000–1100 Law III: legal aspects of robbery, burglary, and trespass

1100–1200 First Aid II: wounds

1200–1300 Lunch

1300–1400 Other Law-Enforcement Agencies: guest speaker, port or local police

1400–1500 Hazards: alarm systems, guard response, prevention, and reports

1500–1600 Routine Patrol I: foot patrol

1600 Dismissal

Third Day

0800–0900	Report Writing 1: basic forms and techniques
0900–1000	Other Law-Enforcement Agencies: Customs-guest speakers
1000–1100	First Aid III: fractures, sprains, strains, and dislocations.
1100–1200	Law IV: use of physical force and deadly physical force
1200–1300	Lunch
1300–1400	Report Writing 2: the complete narrative
1400–1500	Ethics
1500–1600	Cargo Documents I
1600	Dismissal

Fourth Day

0800–0900	First Aid IV: strokes and heart attacks
0900–1100	Law V and VI: other crimes
1100–1200	Law VII: rape, rape evidence, and handling the victim
1200–1300	Lunch
1300–1400	Other Law-Enforcement Agencies: Postal Security and Immigration-guest speakers
1400–1500	Hazards: chemistry of fire and principles of extinguishments and rescue
1500–1600	Cargo Documents II
1600	Dismissal

Fifth Day

0800–0900	Routine Patrol: car and selective patrol
0900–1000	Radio Codes and Use
1000–1100	Intelligence and Organized Crime
1100–1200	Courtesy
1200–1300	Lunch

1300–1400	First Aid V: artificial respiration
1400–1500	Cargo Documents III
1500–1600	Labor Problems
1600	Dismissal

Week Two

First Day

0800–0900	Safeguard Crime Scenes and Evidence
0900–1000	Self-Defense
1000–1100	First Aid Examinations
1100–1200	Traffic Direction
1200–1300	Lunch
1300–1400	Lost Property
1400–1500	How to Accept Supervision
1500–1600	Sociology I: perception and empathy
1600	Dismissal

Second Day

0800–0900	Sociology II: crisis intervention
0900–1000	Sociology III: handling mentally disturbed persons
1000–1100	Hazardous Cargo: safety precautions and emergency response
1100–1200	Sociology IV: minority groups and interpersonal relations
1200–1300	Lunch
1300–1500	Law VIII: arrest, search, seizure, and review
1500–1600	Law Test
1600	Dismissal

Third Day

0800–0900	Final Test
0900–1200	Tour of Port and Its Security Systems
1200–1300	Lunch
1300–1500	General Review
1500–1600	Graduation
1600	Dismissal

PRECAUTIONS TO BE TAKEN BY VESSELS TO AVOID ATTACK BY TERRORISTS

Masters, owners, operators, and agents should be encouraged or required to implement such precautions as are reasonable and appropriate to the situation.

1. Establish liaison with the captain of the port.
2. Establish a procedure for screening own employees, both those who sail and those who service the vessel in port.
3. Hire additional *uniformed* armed security officers and plain-clothed security personnel.
4. Arrange for local police to be visible in area.
5. Conduct pre-sailing underwater body check for limpets with contract diving service.
6. Conduct pre-sailing inspection of all vulnerable spaces and equipment, such as engineering spaces, ship-control spaces, electronics spaces. Look for bomb plants in places that could disable the vessel, such as thrust bearing, safety valves, voltage regulators, steering machinery, HP steam lines, main injection sea valves, etc.
7. Establish fast communications procedures with local authorities.
8. Install secondary communications capability.
9. Limit ingress to vessel to passengers, crew, and those on official business.
10. Establish electronic screening of boarding personnel.

11. Search all hand-carried luggage as is done at airports.
12. Establish electronic or X-ray monitoring or visual inspection of passenger baggage and all linens, commissary supplies, ship stores, and cargo.
13. Install closed-circuit TV in appropriate spaces.
14. Where possible, insist that casual crew (entertainers and the like, other than the mariners) be bonded by a reputable bonding agent.
15. Check for stowaways.
16. Uncover lifeboats, and keep them uncovered until this concern subsides.
17. Designate a security officer.
18. Lock, with controlled lock system, all doors not necessary for use by crew or passengers, and keep them locked. Such doors would lead to stowage spaces, voids, machinery rooms, etc. make an individual personally responsible for the lock series, and have a master key control set up under a designated officer. In a large ship this will probably have to be set up by a commercial security specialist.
19. Search all trunks, etc., in luggage compartment before permitting them to be delivered to passengers' staterooms.
20. Ensure that all emergency equipment is in proper working order and up to specified allowance, and that the crew is familiar with its use.
21. Prohibit vendors from boarding the ship.
22. Spot-check cargo as time and availability of personnel permit.
23. Examine all passenger passports at the time of boarding.
24. Make it known at the time tickets are sold that the presentation of passports will be necessary to board the vessel.
25. Remove all visual identification markings from ratio room exterior so as not to make area conspicuous.
26. Access to compartments should be limited to specifically authorized crew members.

27. Establish ship's watch for small-boat activity and passing of any gear to passengers via any mode from outside the vessel (small lines, throwing objects, etc.). Patrol the vessel with armed crew in small boat under extreme risk or actual threat.

MARINE SECURITY QUESTIONNAIRE

Date: ____________________

Time: ____________________

Firm: ______________________________ Phone: ____________________

Location: ____

____

Person Interviewed: __________________ Title: ____________________

Address: ___________________________ Phone: ____________________

____

____

Description of Complex: ____

____

____

Fenced in? ______________ Yes ______________ No

Hours of Operations: ____

____

____

Type of Commodity Handled: ______________________________________

____

1. PERSONNEL

Security/Criminal Check? _____________ Yes _____________ No

Explain: ____

Credit Check? _____________ Yes _____________ No

Bonded? _____________ Yes _____________ No

By Whom? ____

Total Number of Employees: ________________________________

Union Affiliation: ____

Union Representatives: _____________________________________

____

____

Labor Problems? ___________ Yes ___________ No

Explain: ____

____

____

____

2. SECURITY

How many officers? ___________ Tour Breakdown: ___________ 8–4

Armed? _____________ Yes _____________ No _____________ 4–12

Explain: ____ 12–8

____

Peace Officers? Pay Scale ____________

____________ Yes ____________ No Uniformed? ______________

Mobile Patrol? ____________ Yes ____________ No

Explain: ____

____

Name of Guard Service: ____

Head of Security: ___________________________ Phone ____________

Address: ____

Whom to contact after hours: ___________________________________

____

____

Name of Roundsman: ____

Registered with Waterfront Commission? _________ Yes _________ No

Powers of Arrest: ____

Apprehensions/Arrests Past Experience: __________________________

____

____

____

ALARM SYSTEM

	Central Alarm	____________	Yes	____________	No
Electric Eye	N.D.T.	____________	Yes	____________	No
Push Button	A.D.T.	____________	Yes	____________	No
Closed Circuit TV	Electro	____________	Yes	____________	No
Other	Telephone Co.	____________	Yes	____________	No
	Other ________________________________				

Remarks: ____

____

Types of Communication: ____

____

Monitored by; ____

Who responds: ____

Agencies notified: ____

KEY CONTROL

Administered by: ____

Who has possession of keys? ____

____

Who logs keys in and out? ____

Mastered blanks? ____________________ Where kept? ____________________

____

Grand Masters: How many? ________ In whose possession? ________

____

TRUCKING (Pickups & Deliveries)

Carriers: ____

____

____

____

Bill of lading/Manifest accompany shipment? ____________________

Explain: ____

____

____

How does pick-up work? ____

____

____

How does drop-off work? ____

____

____

Any hold lines? ____

How are trucks controlled? _______________________________________

Who supervises loading/unloading? ________________________________

____

Peak periods of activity: ______________________

How many trucks in and out? ________ Yes ________ No

Procedure: ______________________

____

____

____

____

JANITORIAL SERVICE

Contractor: ______________________

Address: ______________________

Frequency service performed: ______________________

Areas serviced: ______________________

Number of employees: ______________________

Related crimes: ______________________

____

____

____

PARKING FACILITIES (Employees & Visitors)

Who is authorized? ______________________

____

Location of Employee Parking ______________________________

Location of Visitors' Parking ______________________________

Proximity to work area ______________________________

____

____

Proximity to entrances/exits ______________________________

Security of Parking Areas:	Fenced in?	______ Yes ______ No
	Guard?	______ Yes ______ No
	Floodlights?	______ Yes ______ No
Parking Identification:	Decal?	______ Yes ______ No
	Permit?	______ Yes ______ No
	Other?	______ Yes ______ No

Explain: ______________________________________

____

Is there a buffer zone between loading area/cargo and parking? ______

Vehicle parking? ______________________________

Explain: ______________________________________

____

____

____

TENANT PROBLEMS AND COMPLAINTS

____

____

____

____

____

____

TENANTS SUGGESTED MEANS OF ALLEVIATION

____

____

____

____

Interviewers: ____________________

INTERVIEWERS' CONFIDENTIAL COMMENTS:

____

____

____

____

____

____

Part Four

A RAILROAD PERSPECTIVE

A RAILROAD PERSPECTIVE

CARGO THEFT PREVENTION FOR RAILROADS

The development of a cargo theft prevention program involves the cooperation of all who are even remotely involved in the movement of freight. It is only fair to confess that what follows is written from a common carrier's perspective, more specifically, written from a railroad viewpoint. However, it is felt that the principles and guidelines that are outlined in this chapter can and should be applied by all who are involved in the transportation cycle.

WHAT IS YOUR PRESENT CONDITION?

No matter what the mode of logistics is, before the improvement process begins, you have to have a good understanding of your present condition. The reason is simple: you have to know where you are before you can determine where you are going. This point can best be illustrated by the various weight-control and diet programs that abound throughout the country. The selling points center around this concept. They carefully measure your present deteriorated condition of obesity, making you aware of every negative detail. They use your present condition to convince you that it is time to improve the situation by subscribing to their program. During the improvement process you are again closely monitored until your goal of weight loss is reached. No matter which program is involved, the selling points are always the same and centered around the classic "before" and "after" pictures, contrasting a 293-pound blimp as opposed to the same person one year later wearing a size eight bathing suit and weighing in at a svelte 105-pounds. The sales method is obvious but effective.

Though this guideline may be familiar to most of us, and can hardly be considered original, it is unfortunate to note that the transportation industry as a whole continues to have trouble getting a handle on their present condition in regards to cargo theft.

INVOLVING EVERYONE CONCERNED

The following fictional story is typical of what takes place in the transportation industry on a regular basis.

A pair of size 14 shoes are manufactured by the X-L Shoe Company, located just outside of Taipei, Taiwan. The shoes, along with eleven other pairs, are boxed and placed in a carton. The carton is placed in a truck trailer belonging to Chu Ling Trucking Company and drayed to Far East Freight Consolidators in Taipei. The carton is then placed in an intermodal container. The container is transported to the Port of Taipei, again by Chu Ling Trucking. At that point the container is placed on a container ship bound for the Port of Los Angeles. After unloading at Los Angeles, a container is hauled by Malibu Trucking to the local railroad piggyback yard and placed on a rail flatcar. The container moves on a train from Los Angeles to Buffalo, New York. In the process, the car is handled by five different railroads. From the Buffalo TOFC ramp, the container is hauled by Upstate Drayage to a distribution warehouse in downtown Buffalo. The container is unloaded and our shoe carton is placed in a city delivery truck destined for Gowanda, New York, located just outside Buffalo. The carton is finally delivered to the Erie Department Store in Gowanda, where the size fourteen shoes will be sold to Joe Huch, who had special-ordered the shoes six months earlier.

When you contemplate this rather common occurrence from a cargo-security standpoint, three major facts stand out: (1) a surprisingly large number of people are involved in the movement of cargo; (2) given the number of entities involved, it is a credit to our transportation system that in the majority of cases cargo arrives on time and intact; and (3) if anything is stolen during the transportation process, it will be extremely difficult to identify and resolve the circumstances that contributed to that theft.

Suppose for a minute that when the carton arrived in Gowanda, it had been torn open and our size 14 shoes were missing. In addition, the last 10 shoe shipments from Taiwan to Gowanda had experienced similar pilferage without any explanation. Also, suppose that you had been given the charge to resolve this chronic pilferage problem and prevent it from happening in the future. What would be your first move? In most criminal investigations a good place to start would be to interview the victim in order to get the facts of the case. In our shoe caper just who is the victim? In this imaginary case, all sixteen transportation entities would be the victims. Either directly or indirectly, all members of any transportation chain will pay for any loss of cargo that occurs between origin and destination. Therein lies a concept that most transportation companies fail to consider when designing their theft-prevention efforts.

Almost without exception most transportation companies limit their theft-prevention efforts to the physical boundaries that company controls. Little, if any, thought and effort are given to theft prevention outside those limits. On the surface that approach makes sense; after all, it is human nature to protect one's own interests first. But remember, the transportation industry is a very complicated inter-dependent entity and some of the standard rules just do not apply.

Our case involving the shoes would be a good example. If all 16 companies that handled the shipment were made aware of a chronic pilferage problem, they would all immediately scramble to show that the thefts did not occur while the cargo was in their control, but in all probability little communication would take place among the 16 companies concerning what could be done to jointly prevent the problem from happening in the future. It would simply be a case of every man for himself. (It is interesting to note that this attitude of taking care of "No. 1" was also a major obstacle in the DOT's effort to formulate a national cargo-security program). Many would argue that you should take care of your own corporate interests first, and that such an approach does produce results. When listening to those arguments, one again thinks back to the Platte River Project approach, and wonders what kind of results are really being produced. It would be my opinion that by going it alone, a single corporate entity will produce only short-term, superficial results.

THE CUSTOMER IS KING

The shipper is, in fact, the kingpin in the whole theft-prevention effort, and it is that entity out of all who are involved in the transportation chain that will determine the success or demise of your efforts. If the customer is totally involved in the prevention process from planning through implementation, then you will have a much better chance of succeeding. It is a simple case of power and influence. I am reminded of a very humorous situation that portrays the power a customer can wield in a prevention program. For obvious reasons, the specifics of who was involved shall remain anonymous.

A customer with whom we had been working for some time suddenly began having serious theft problems with shipments for one of his customers on the East Coast. Because of our relationship in the past, our railroad was immediately contacted by this customer to help resolve the problem. The problem involved shipments that originated in the West and moved to an East Coast destination. These shipments moved over three different railroads en route to their destination. Working with the customer, we contacted the other two railroads involved in the move. One of the railroads was very helpful and wanted to assist in every way possible to resolve the problem. The other railroad, for various reasons, was somewhat reluctant and hesitated in making the appropriate commitments needed to proceed. Its hesitance greatly hindered the whole process. The customer, who was not used to this type of halting response, made appropriate contacts, which might be interpreted by some to be threats, with rather high-level officers within that hesitating railroad's power structure. Literally within hours, the hesitant railroad had totally changed its outlook on life. Now, instead of holding back, it suddenly became very supportive and could not do enough to speedily bring the problem to a resolution.

In a very vivid manner this story illustrates the power a customer can have in expediting the prevention process. Often he is the only one who can speak with authority to everyone who is involved with the movement of his product. Just as important, he generally has the power to exclude any weak link from the chain of transportation that moves his product, particularly if that link is uncooperative.

The kind of power use outlined in this example obviously should not be considered the best way to utilize the customer's influence. In fact, this "hammer-and-tong" method should be used only as a last resort, and even then it should be limited. The important point to consider is that the customer's support, even if only implied, is essential to your effort. It will open doors that could not be opened through any other source.

Corollary

If it were not for the many mistakes that are often made by those who would utilize the shipper's influence in their prevention programs, it would be wise not to mention what follows, because it would seem to be a natural consequence of the customer-involvement concept. Unfortunately, the natural consequence does not always follow, hence the need to briefly cover this important point.

Because your whole purpose in involving the customer is to marshal his power and support to implement prevention techniques, it only makes sense to make that contact at the highest possible level in that corporate management structure. The contact has to be at a level where corporate policy is formulated and placed into action. We have learned through sad experience that that level of influence is certainly not at the warehouse door, at the divisional level, or even at the regional level, but invariably lies at corporate headquarters.

Though there are many new-wave management philosophies that would try to convince you to the contrary, the facts of the matter are that company policy is still made at corporate headquarters, and that power still flows from the top down. You cannot gain the total support of a company for your prevention program by convincing the warehouse foreman and regional claims manager, and hoping that they will pass the word upstairs. It would be like an insurance salesman trying to sell group life insurance to the company by trying to convince each individual employee of the policy merits. The decision is still made at the top. This philosophy should in no way detract from the importance of involving the middle and lower management ranks in the prevention effort, because they are critical to the implementation process. However, implementation should come as a result of corporate policy emanating from the top rather than from fragmented localized efforts initiated by an outside entity that could never get within shouting distance of the corporate boardroom.

LOCATING THE PROBLEM

Suppose that you have worked hard and have developed what you believe to be accurate data sources. Your cargo-loss figures are valid and you feel that you have an accurate gauge concerning the magnitude of your problem. In addition, you have been effective in enlisting the aid and support of all those involved in the movement of a particular problem shipment. The next big question is, Just what are you going to do next to locate the point(s) in the transportation process where your cargo is disappearing? In simple terms, who is stealing your cargo and where are they stealing it? As promised earlier, no attempt will be made to give a detailed outline of methods that may be used to locate the points in the transportation process where cargo is particularly vulnerable to theft. However, a few general concepts do apply in most cases, and they deserve mention.

At the risk of being criticized for being prejudiced because of my common-carrier background, may I suggest that the bulk of cargo theft does not take place during the actual movement of the freight; that is, while the cargo is in or on a transportation vehicle in transit from point A to point B. Despite highly publicized accounts of truck hijackings, systemized looting of railroad boxcars, and even recent accounts of increased piracy on the high seas, the majority of cargo loss does not occur en route.

Most in the transportation industry would agree with this statement, and in support of that view, they would quote you authoritative data in which the majority of cargo loss occurs during the loading and unloading processes. Certainly, personal experience would tend to support such a view, but before you relegate the carriers to a secondary role in the prevention process, remember the interdependent nature of the transportation chain. The same people who developed the data mentioned above also brought you the other solid statistics referenced earlier. Suffice it to say, if you are suffering freight shrinkage, chances are that this loss is at someone's shipping and receiving dock, and chances are that you are losing your freight one piece at a time.

Whenever a major theft occurs in the transportation industry such as a hijacking, which is the result of a single event, both the visibility of and the response to that event are overwhelming. The loss quickly becomes a glamour event with everyone from the town marshal to the FBI expressing interest. Even more importantly, your own management will suddenly become very visible, uncomfortably so. They will become very concerned about resolving the problem, making sure that it does not happen again. Management certainly cannot be faulted for their concern. They are justified in their response. Such single events often result in losses that could rival the Great Train Robbery.

Yet it is astonishing to observe that same management calmly accept annual cargo-loss reports that literally run into millions of dollars every year and not so much as bat an eyelash. Why? Because those millions of dollars represent a large number of small thefts, most often committed by someone who works in the transportation chain. Because of his legitimate position, the typical cargo thief is very difficult to detect. The phenomenon that allowed our friend in the department store to walk off with $1,000,000 is also alive and well in the transportation industry!

So, in the majority of cases, your cargo-loss problem will center around those points in the transportation cycle where the cargo is legitimately outside the confines of a transportation vehicle, and it will involve a multitude of small thefts perpetrated by insiders.

Even armed with the knowledge that most of your theft problem will probably take place during the loading and unloading processes, it is still difficult to locate the exact point where the loss is taking place. Besides, just because the majority of theft is assumed to take place at the dock does not eliminate the possibility that your loss might be taking place in transit. The seal applied to the transportation vehicle now becomes paramount to the identification of problem location. This fact will come as no surprise to anyone even remotely associated with the cargo-loss problem. The simple seal concept has remained virtually unchanged for literally hundreds of years; that is, to ensure to all parties concerned that the transportation vehicle has, or has not, been entered en route.

A good example of the role that seals can play is our imaginary shoe shipment described earlier. If the same seal placed on the container in Taiwan was intact when the container arrived in Buffalo, then suddenly the number of possible loss locations is reduced from 16 to four, and you have confirmed that the loss is taking place during the loading and unloading process. The value is apparent.

Until recently, seals and the concept that supports their use have experienced little innovation. However, two very important changes have taken place during the last few years: (1) the recording and maintaining of accurate seal records has fallen off sharply, the obvious factor being the cost of generating and maintaining such records; and (2)the improved ability of the transportation industry to expedite the movement of cargo. That is, freight simply moves too fast through key interchange points to allow complete seal accountability. Again, our shoe shipment clearly outlines these factors. In order to maintain a perfect seal record, it would have been necessary for all 16 entities involved in the shoe movement to make appropriate seal inspections, record those inspections, and store those records for future reference. From a cost-effective standpoint, such exhaustive seal records are not justified. Even if the transportation companies were willing to expend the money and manpower, it would still be difficult to obtain seal records. Because of these realities, in most cases the only seal inspections made, in all probability, will be made at origin and destination.

In response to this lack of seal inspections en route, the basic seal has recently taken on a new dimension. Within the last decade there has been a literal explosion of what can best be described as high-security seals. They come in all shapes and sizes. But they all add a new security dimension to the basic seal concept by making it more difficult for the would-be thief to gain entry to the transportation vehicle. In many cases, the impact of these devices has sharply reduced the number of thefts en route. However, they can hardly be considered a panacea for theft prevention, for the bulk of cargo loss occurs outside the transportation vehicle itself.

SUMMARY

The cargo theft problem has always been difficult to deal with because of the intricacies of the transportation industry itself. The prevention of cargo theft does not lend itself to a simplistic generic solution.

Despite its complicated nature, cargo theft can be prevented via a cost justification standpoint. However, the approach taken must be flexible enough to accommodate the many unique theft problems that the transportation industry presents. Not only is the industry complex, but it tends to be in a constant state of change. As a result, what might be an effective prevention approach today may be rendered invalid tomorrow.

Despite these variances, experience has revealed several general elements that should be an integral part of any effective theft-prevention program. Though at times these elements may appear somewhat simple, it is their exclusion that invariably leads to an ineffective prevention effort. They are as follows.

What Is Your Present Loss Condition?

Failure to accurately diagnose your present loss levels often leads to inappropriate remedies. This factor was the major reason for the failure of efforts to formulate a national cargo-security program during the 1990s.

Involving Everyone Concerned

Transportation is a highly integrated industry. If one member sneezes, another gets a cold. Though it is a natural tendency to protect one's own corporate interests first, you simply cannot solve the problem of cargo theft by limiting yourself to the boundaries of your own corporation. All involved in the movement of a problem commodity must also be involved in the resolution process in order to prevent it from recurring in the future.

The Customer Is King. No other member of the transportation chain has more ability to implement the theft-prevention process than does the entity to which the cargo belongs. Any attempt to develop a program without the approval and support of the customer will be difficult at best. The customers' overt support can also be the key to involving other members of the chain who would otherwise hesitate to become involved.

Corollary. The support of the customer for your prevention program must come from the policy-making level within that corporation. If support is generated at a lower level within the corporate hierarchy, the results will be localized and short term in nature rather than long-term corporate policy.

Locating and Dealing with the Actual Problem. Individual circumstances will dictate what specific methods will be used to locate and deal with specific loss problems. Though a generalized approach is favored, several factors deserve mention.

1. Though not backed up with empirical data, experience has indicated the bulk of cargo loss takes place during the loading and unloading process rather than in transit.

2. Most cargo losses probably represent a large number of small thefts perpetuated by insiders rather than a small number of large thefts.
3. Seals play a critical role in pinpointing the location of cargo theft. Since the introduction of high security seals, seals have taken on a preventive dimension.

CONCLUSION

The information outlined in this section is intended to be helpful to the person who is contemplating the formulation of a cargo theft-prevention program. Despite my inclination to be more specific in addressing the subject, the dynamics of cargo loss would not allow it. However, it is my sincere belief that if the general concepts that have been outlined are properly used in your prevention effort, you will have a program that will accurately address the real causes of cargo theft.

Part Five

SPECIAL AREAS OF TRANSPORTATION SECURITY

SPECIAL AREAS OF TRANSPORTATION SECURITY

CRUDE OIL MOVEMENTS BY TANKER

Crude oil is generally moved in the basic methods of tanker, barge, pipeline, or tank truck. The greater quantity is moved by tanker, which may range from small coastal tankers to very large crude carriers (VLCC), which are sometimes so big that they cannot get into harbors to load or unload. Crude oil can be pumped off or "lightered" into barges that can then be maneuvered closer into shore for unloading. Crude oil is also brought to delivery points from pipelines that can range from small one- or two-inch lines directly from wells through large interstate pipelines that move the oil to refineries. Crude oil can also be moved by tank truck, generally from remote wells to a gathering point when the cost of laying a pipeline is prohibitive. A tank truck can carry approximately 200 barrels of crude oil at any one time.

In large-scale tanker movements, arrangements to purchase or to sell the crude are made in the major trading areas of the world, where the quantity and quality and price are established. The empty tanker is directed to a loading point whose configuration varies greatly according to the terrain, the amount of oil involved, the depth of the water, etc. The loading is supervised generally by a third party, whose duty it is to see that both parties' needs are satisfied; this party is paid a fee for his services. In some cases the tanker is loaded through huge pipelines that pass the oil through meters that tabulate the amount. The meters vary in configuration and have a design accuracy of about 0.10 percent. Generally the third party, along with representatives of both carrier and shipper, physically measures the depth of the oil in the ship's hold by using the *strapping* tables provided when the ship was manufactured. These tables tell the volume of the ship's tanks according to depth of oil measured. The temperature is recorded and a sample of the crude is taken for laboratory analysis.

Upon arrival at the destination the process is reversed, with another third-party representative again measuring the volume of the oil on the vessel along with a sample, when the crude is pumped to short tanks. Sometimes the oil is also metered at this point.

For years this procedure was thought to be a secure way of transporting crude oil because oil was cheap and diverting large volumes of crude from the tanker itself was a difficult task. It was also understood that there would be losses because the *high ends* or the lighter elements of the hydrocarbon liquid would evaporate, leaving some of the heavier crude substances to cling to the sides of the ship (called *clinnage*). In addition, of course, there would be temperature variations and some spillage because of careless handling by members of the crew. After the price of crude rose dramatically, there was a parallel move to limit the permissible amount of evaporated hydrocarbon. In order to protect the quality of the air we breathe, laws were also enacted to prohibit the spilling of oil on coastal waterways. One could expect that the amount of losses suffered by a receiver would diminish. To protect against evaporation, the hatches on many tankers were sealed and the cargo was literally flooded with inert gas. New handling techniques were developed along with better valves and drip pans to preclude the possibility of spillage of the crude. Some vessels, especially those that carry heavy crude, use heating coils that further liquefy the heavy crude so that it can be pumped off more efficiently, leaving virtually none clinging to the sides of the ship's tanks.

Diverting Crude Oil

Despite these improvements, the value of the crude oil still creates temptation, and methods have been developed to divert crude oil. Bribery and corruption are not unknown in those areas of the world where crude oil is produced and shipped, and sometimes false certifications of loading are discovered. Because many oil-shipping terminals in foreign countries can be arms of their governments, it is sometimes difficult to establish a proper redress should this occur. When this type of activity is suspected, very sensitive situations develop that require extensive treaties involving international ethics and diplomacy.

One way of diverting crude oil that has been detected on several occasions is to pump off crude oil as *slop oil*, which is then subtracted from the ship's cargo. This slop oil may have the same value as the original shipment and is sold for the profit of the ship's crew. A 2,000-barrel diversion of a 300,000-barrel shipment is less than one percent but equal to thousands in tax-free profits. It depends on the market price.

Some oil is moved from port to port by using *spot charters*, which are tankers that move around looking for fares, much like taxis in the urban areas of the United States. Shipments using spot charters are generally at greater risk because the temptations become greater, the controls become fewer, and the possibility of redress becomes more remote because of international ramifications. Ship captains have been known to remove some high-quality crude and cover the liquid loss by adding emulsified water to the cargo. This would possibly be detected upon laboratory examination, but the tanker may have unloaded and be out to sea before the deed was discovered. Ship captains have also been known to present their fuel charges to the consignee or lay claim from the ship's owner for fuel costs in cash because of monetary problems throughout the world. Some captains have installed devices called *nopals* that allow crude oil to be burned directly by the ship's engines. The nopals are generally installed in unethical shipyards and are disguised as plumbing repairs. The ship captain can then complete his voyage, burning the cargo yet submitting a phony invoice for fuel charges that is then paid in cash. Although competent maritime engineers claim that burning of crude oil in diesel and steamship engines is detrimental to the engines and is unsafe, it is considered by others to be one means of getting rich quickly.

Crude oil has also been diverted on a large scale by simply unloading the entire cargo in an unauthorized port and sinking the tanker, but this is generally an insurance fraud and although very costly, it should be treated elsewhere.

CRUDE OIL MOVEMENT BY BARGE

Barge movements parallel tanker operation in examples of threats to the integrity of the cargo. Barges are generally used on coastal waterways to move medium cargoes by being either pulled or pushed by tugs. Some barges are sometimes left loaded like movable storage tanks. Unauthorized changing of the strapping values of barges has been known to happen in some areas of the world. When the barge is initially constructed, the volume of its tanks are carefully calculated and become almost unique to the vessel.

Diverting Crude Oil

Unscrupulous parties have found ways to deceive the petroleum measurement inspectors, who drop a measuring tape through the crude oil to get the exact depth. They have been known to put objects on the bottoms of the tanks to disguise the true depth of the oil that is put on the barge, therefore giving the crew the ability to carry crude oil "over without billing," subject to resale for their own profit. In some configurations, barges are unloaded miles from the receiving tanks that are used to store and measure the amount of delivery.

For various reasons, a pipeline would be placed between the barge and the final destination of the crude; this pipeline can be tapped by unauthorized parties. This sort of tapping is rare but it can be very costly; it will be described later in more detail.

METERS

Determining the total volume of crude oil by metering is quite often done by moving a measured volume of oil through a calibrated point during a set period of time. Meters made of various configurations perform this function, and within limitations they are used throughout the world to measure moderate volumes of oil efficiently. Meter size ranges from small meters used on oil leases, known as *lease* or LACTs, to large meters that fit on 10-inch or larger lines usually for ship movements. Meters are read directly or indirectly depending on the need, and they are generally designed without any thought given to protection against tampering. Meters are now being linked to computer systems that automatically feed their data into machines for tabulation and billing.

Under the auspices of the American Petroleum Institute (API), guidelines have been established that were designed to protect the rights of both shipper and receiver by measuring the product through an approved metering system. Because a meter is generally a permanent connection, both parties may agree to the purchase and maintenance of the meter, but they also must agree on the *meter proving* operations. Meters, being mechanical devices, are subject to variation and fluctuation and, therefore, they are proved regularly. Proving is generally done by a third party who has specialized equipment to send a given amount of product through the meter at a given temperature; careful adjustments are made to get the appropriate measurement.

The adjustments are tabulated and agreed to by both parties involved, and the meter is then sealed until the next proving episode takes place. Should the meter malfunction between proves, a meter factor has been established and the differences involved are generally halved so that neither party can be effectively cheated.

As in any business, meters, being the court of last resort in the transfer of product, can be altered to benefit either the buyer or the seller to the detriment of the other. Meter provers can be bribed to give an unjust percentage to one party. In some cases, the meters themselves can be adjusted, retarded, or the numbers increased without leaving a trace. The techniques involved depend on the type of meter, mode of installation, built-in protection, or other factors. *It should be remembered that meters were never intended to be tamper resistant.* One need only to remember that when they were first introduced to our society, vending machines were also very vulnerable to tampering. Many dollars were lost to the vending industry until virtually tamper- and vandal-proof machines were developed and deployed.

Meters have also been bypassed by adding a separate pipe around the meter to allow more product to enter into the system without being recorded, or to allow product to flow back into the shipping system while billing the customer for product never received. Close examination of the plumbing surrounding meters is always in order. Some bypasses are necessary for safe operation, but good security practice would dictate that these bypasses have valves that are normally kept closed and sealed and protected from tampering. It is again not unknown for the valve flapper to be drilled prior to its installation by unethical people, to allow oil to pass through the valve although the valve indicator would register closed.

Meter and Valve Seals

The sealing of valves and meters in the petroleum industry is always a very important job, but sometimes this job is neglected. The standard sealing procedure is to developing a non-reproducible seal, which is duly recorded and installed correctly, subject to be broken only by authorized individuals. Some devices commonly deployed throughout the industry used for sealing are routinely compromised by knowledgeable people. Great care must be used in inspecting any "sealed" system because what appears to the eye as being adequately protected may just be a sham.

SAMPLERS

In some areas where crude is shipped and recorded by meter, *samplers* are deployed. These mechanisms allow technicians to collect a given quantity of oil for analysis, which helps in determining the value of the shipment. There are both automatic and manual samplers, with the automatic system generally designed to sample one cubic centimeter of crude oil per barrel transmitted through the line. This sample is deposited in a sealed container so that when it is retrieved, a technician can analyze it for contaminants such as salt water. It is very important that the sampler be hermetically sealed as to protect against loss of high ends, the most valuable part of the crude oil, which have a tendency to evaporate most readily. The sampler should also be protected against tampering by locks and seals. Again, caution must be shown because samplers can be convoluted and complex, and values can be altered by the addition or removal of a contaminant from the sample itself.

Some samplers are devices that take only one segment of a shipment. They are merely taps on the line from which a quart or so of product is removed for analysis. This is not a preferred sampling technique, for it allows for only a small actual sample of a large amount of oil passing through the line, and only at that moment when the sample is taken. It does not allow for sampling a continuous run of crude, which may vary in quality. A variance of one or two percent in the quality of a crude shipment may mean a considerable amount of money to either buyer or the seller, who stands to benefit by the variance.

Samplers at Oil-Well Sites

On *leases* or oil-well sites, samplers are often used in conjunction with the lease automatic custody transfer unit. The oil is pumped from the ground and held in a holding tank until, at a predetermined time, it is sent through the meter and through the sampler to a pipeline to a gathering system. Because custody is changed at this point, or a sale is made, both quality and quantity must be established. The custody transfer unit can be clock controlled and can provide for adjustment of the quality of the oil that can be put into the pipeline at any one time. Oil coming from the ground sometimes varies in the amount of salt water it contains, and methods are usually in place to remove the salt water. Should the sampler detect a large amount of salt water, for example in excess of two percent, the oil will be sent through the water-removing process and then back into the holding tank. The oil will continue to circulate through the water-removing process until the sampler automatically detects an amount of water in the crude oil of less than two percent, at which time it will allow the oil to be shipped into the pipeline.

Unscrupulous leaseholders or producers have been known to defeat the sampler or the timer by simple mechanical or highly complex means. Some samplers have been defeated by merely drilling a hole into the top of the cabinet so that the acceptable ratio of water to oil can be adjusted improperly. Again, care should be taken to fully inspect the entire sampling and transfer system so that tampering cannot readily take place. The control cabinets, power supplies, relief valves, and all other points of entry or control for the custody transfer unit should be thoroughly locked and sealed, subject to routine inspection. Care should also be taken to prohibit the installation of a bypass or any type of backflow device that would allow the crude to go back into the producing tank after it has passed through the recording meter.

MOVEMENT OF CRUDE OIL BY TANK TRUCK

The final method of moving loads of crude oil is by tank truck. Generally, a tank truck is dispatched to a lease when there is a sufficient amount of crude oil to be removed and sent to a gathering point for distribution. Many lease tanks have valves located near the bottom, which when connected to a hose from the tank truck allow the tank truck to be loaded either through pump or gravity means. The tank truck can hold approximately 200 barrels and delivers sometimes through a *receipt meter* located on a pipeline. Oil costing $35 a barrel readily raises the theft potential when controls are weak. The producer must be careful to keep accurate well records to determine the amount of oil produced by his well at a given period of time in addition to the amount given to the trucking company to haul to the pipeline meter. In some states, and in foreign countries, it is quite common to have convenient locations where crude oil can be readily sold and converted into cash without question. In some states, these points are called *reclaimers*; their operation is to buy up various types of crude oil from tank bottoms, slop pits, or other minor sources of production; process them slightly; and then sell them to a pipeline as crude oil. States like Texas regulate the reclaimers and their theft potential minimal, but other states have no such control, allowing for easy "fencing" of stolen oil. Thefts from leases are often conducted at night by small tank wagons, sometimes operating with a large "mother" tank wagon, and they may go from lease to lease, stealing 10 to 20 barrels per stop. This oil is then moved through the illicit market place and converted to cash. Nighttime oil thefts are readily detected when good records are kept and the lease is kept clean.

Care must also be exercised in reviewing delivery tickets, run tickets, bills of lading, or other documents reflecting the production and movement of crude oil. Frauds conducted in this manner, or when counterfeit or falsely constructed documents are used to "prove" delivery of crude, are not unknown. A huge quantity of oil available in a pipeline system at any one time makes it unlikely that small differences will be readily detected. Even after these differences have been detected, it is difficult to determine where the loss occurred.

Diverting Crude Oil

Because of the huge amounts of crude oil, it is also possible to disguise shipments of salt water or other contaminants as crude oil without readily being detected. Fluids of any kind can be used to move meters. When the product is not appropriately sampled, these fluids are used to generate revenue for the unscrupulous trucker. On several occasions major refineries have suffered large losses when the catalyst used to refine the crude oil becomes contaminated by chlorinated solvents fraudulently introduced into a pipeline system in an attempt to "beat the meter." Environmental rulings have prohibited practices such as dumping salt water on the soil and, therefore, service industries hauling salt water to injection wells have developed. Unscrupulous drivers have been known to inject salt water into the pipeline systems in an attempt to get rid of their cargo and yet obtain the proceeds of a delivery of crude oil into a system. Control over the ability to enter the pipeline system and a tamper-resistant sampler at every point of entry should be mandatory in any area of moving crude oil.

REFINED PETROLEUM PRODUCTS

As crude oil moves through the system, whether from barges, tankers, or pipelines, it ultimately will end in a *refinery*, where it is converted by a series of complex processes into various hydrocarbon products.

The products can consist of everything from gasoline to lubricants, from feedstock to plastics and medicines to asphalt, sulfur, and coke. The type of products depend on the kind of process used and the type of crude available to be processes. Once crude oil is in the system, it is relatively safe from pilfering. It is then converted to a useful product of considerable value and generally of interest to the regulatory authorities for taxing purposes.

Once produced, the products are moved in various ways, by pipeline, tank car, barge, or tank truck, to the point where the customer can utilize the product. Gasoline is sent through pipelines to terminals and ultimately to service stations; feedstocks are moved by pipeline or tank car to chemical plants or pharmaceutical companies, where they are converted into useful products. Somewhat unique to the petroleum industry is the fact that products, though having different names, may have an overlap in specification tolerances. It is therefore possible to have types of kerosene that may qualify for jet fuel, or for diesel fuel, or for home heating fuel. The specifications of the ultimate customer are important, and additives are included that increase the performance of the product in its intended use. It is, therefore, theoretically possible to burn jet fuel in a diesel truck or in a home oil burner, although to do so would not be financially sound. It is also possible to burn various grades of gasoline in automobile engines, but using an inferior gas may cause damage to the engine over long-term use. Certain feedstocks—highly refined hydrocarbons—would burn in an automobile engine, but they would be far too costly and would probably cause long-term detrimental side effects. Specifications for asphalt are similar to those for number 6 bunker fuel, with asphalt receiving a hardening agent before being applied on a road for pavement. It is this phenomenon that allows for the easy conversion of certain fuels in the underground marketplace. A tank wagon of feedstock can easily be dumped into the ground tanks of an unethical service station, commingled with the existing product, and sold as regular gasoline without the customer becoming suspicious. The service-station operator would not be concerned that the feedstock is very highly refined and not intended for automobile use because he did not have to pay its true value for it.

Because refined petroleum products have increased dramatically in value, their theft potential has also been increased. In many cases, the physical controls used to monitor the movement of the product have not been refitted to reflect the increased security hazard. Environmental procedures such as bottom loading and unloading and vapor-recovery systems have eliminated the argument that the product may have evaporated or been spilled during transfer. A loss allowance of 0.5 percent again was traditional in moving the fuels, but this is gradually being eliminated by the use of contracts and tariff regulations that specify the exact quantities at a standard temperature and pressure. Therefore, no loss is now permissible.

Obviously, a loss can involve a serious safety hazard, jeopardizing the existence of an entire load and the safety of the individuals around it.

MOVEMENT OF THE REFINED PRODUCTS BY PIPELINE

Over the years, the development of the gasoline-automotive business in the United States has laced the country with pipelines delivering the product to the marketplace. Pipelines are the most efficient way of moving the refined product, and although it is done only rarely, taps have been placed to allow for the unauthorized removal of the product. Generally, pipelines are placed underground for corrosion resistance and for physical protection, but there have been several successful incidents where unscrupulous people have removed quantities of product over a period of time, thus enriching themselves unlawfully. Because of variations in the temperatures, pressures, and volumes involved, a small percentage of loss is sometimes undetectable or is explained as being the result of the compressibility of the fluid. Most pipeline operators have specialized equipment to tell if a major break has occurred. Although this equipment automatically shuts down pumps, a tap may allow a small percentage to trickle away without readily being detected. Devices, known as *pigs,* are routinely sent through pipelines to determine the amount of corrosion and of any other abnormalities that may develop into the system. These are also known to detect large metal taps, but small taps may pass their scrutiny. Tapping a pipeline full of flammable liquids is always a very dangerous operation and sometimes results in the destruction of the tapper and his equipment.

A pipeline functions much like a railroad, for it transfers the product by switching from line to line to move the product to the intended destination. At points of custody transfer meters are provided along with devices to determine the type of product in a pipeline. The meters are operated quite similarly to other meters, and they are subject to frequent proving to the satisfaction of both parties to a transaction. The pipelines then feed into tanks located near terminals where the product is ultimately going to be distributed to the customer, generally by tank wagon or delivery truck. These trucks deliver directly to gas stations, fuel oil customers, or major accounts such as utilities, airlines, or trucking companies. These meters also are routinely proved and should be subject to routine sealing and inspection by both parties to any transaction.

CONCLUSION

As in most areas of business, sound business practices, if rigidly followed, would almost eliminate the potential for theft in interstate commerce. When human error, inattention, and the pressures of business modify sound business practices, losses can and do occur with great regularity. Awareness of the threat potential, firm adherence to basic cargo-security techniques, and common sense will go a long way in precluding theft-related losses in the petroleum industry.

TRANSPORTATION OF HIGH-VALUE SHIPMENTS

The protection of "valuables" is perhaps one of the greatest challenges facing the transportation industry today. In this chapter I will attempt to cover various aspects of high-value commodity protection for commodities that span the spectrum from cash to computers, from silicon to securities. I will do this with the goal of presenting ideas that will help you to develop a high-value commodity protection program that is tailored to the needs of your organization.

To deal effectively with the problem of protecting high-value commodities, one must first analytically determine what is and what is not a high-value item. Sometimes this can be quite easy. Obviously, items such as cash, negotiable securities, gems, and precious metals have high true cash values and thus require increased attention. But the determination of what is and what is not high value must also take into account a number of other factors. These include an item's vulnerability or desirability, claim costs, and business impact.

Many items become high value due to their popularity or desirability at a particular point in time. Items such as citizens band (CB) radios, miniature calculators, watches, and computers, usually go through a period of heightened popularity that places them in great demand.

This problem normally is most noticeable around the Christmas holiday period. Such items thus become "target" items, requiring special precautions.

Other items, such as pharmaceuticals and technology items, are often shipped with a "released valuation" or insured value that is much less than their actual market value. Such items are often overlooked if only their released valuation is considered, yet their loss can have a financial impact that greatly exceeds this surface value.

Another important consideration in determining an item's true value is what impact an item's loss may have on a business. The loss of many items can have far-reaching effects. These effects may include a loss of business prestige or reputation, curtailment of normal business operations, an increase in experience-based insurance costs, and many others. This value category catches many businesses by surprise if they fail to properly account for the business value of a new product design, important part, computer tape, etc. Carefully ask yourself what impact the loss of any particular item may have on your business.

Once the value of an item is determined, a decision must then be made on how that item should be protected. In some cases acceptance of the risk without action, or possibly with contingency planning, may be chosen. Most often some form of enhanced security will be deemed appropriate to reduce exposure to a loss. Careful weighing of all factors involved, including the human factors, will hopefully lead to your making a correct decision.

If a determination is made to afford a commodity enhanced security, various means of accomplishing this goal must then be considered. Some of the most basic areas that will require consideration include method of storage, method of transportation, and types of documentation and control. A proper mix of these security enhancements should result in a minimum risk of loss.

STORAGE

Methods of storing items of extraordinary value are dependent upon the physical nature of the item itself, the item's determined value, and various operational considerations. From these factors it is possible to select an appropriate physical storage area, determine the type of security that the area will require (alarms and/or guards), and which personnel will have access to, and responsibility for, the area.

Physical storage areas can vary from a locked desk drawer to a bank-type vault with a wide variety of suitable structures in between. First, you must determine the physical size of your items, and project what will be the maximum number of items you will have to store. From this you can then determine the physical size of your storage area. Once the required physical size of your storage facility is determined, the type of construction must be specified. For small items, a locked desk drawer or an office-type safe may be appropriate. For large items, an enclosed security cage or perhaps even a bank-type vault might be required. Construction specifications will be highly dependent upon the value of the merchandise you wish to protect.

For general-purpose storage of theft-susceptible, but not extremely high-value merchandise, a security cage is often appropriate. These cages are fairly easy to construct in virtually any available space. They typically consist of a cage enclosed in "security fence" material, a high-strength version of chain link, with a suitable gate for entry. With such a cage, care must be taken to ensure that the gate is properly mounted, because it is often your most vulnerable point. Sliding gates should slide on railing inside the enclosure, not outside.

Remember that your security storage area is actually six-sided, not just four-sided. The top and the bottom are of great importance, and are often overlooked. Security for these two "sides" should thus be planned in your initial design. Floors can be reinforced with steel plate if they are vulnerable to penetration, and the top can be covered with security fence if needed to prevent climb-overs.

If your material is of extraordinary value and you will need a safe or vault to store it, check with a number of different manufacturers, and have them explain the features and penetration resistance of their products. For safes, a good reference specification is the TL15 rating. For vaults, 5R is a medium grade. Ask the manufacturers to explain these ratings to you, then determine if you will need more or less protection and adjust your specifications accordingly.

If you find that you need a small storage vault, you may wish to consider one of the new modular units now available. All six panels are precast, then assembled and welded on site. This can result in considerable savings in time and money, and some of the units are now UL rated. In any case, once you conceptualized your storage facility, check with others in your industry to see what they have done. You will in all probability learn a great deal of information, both positive and negative, that will be of benefit to you. Most of all, learn from their mistakes. As someone once said, "You'll probably never live long enough to make them all yourself."

Depending on the nature of your commodity, you may find it cost justifiable to utilize an alarm system or guards for your facility. Both can normally serve the same function—detection and prevention of any attempts to gain unauthorized entry. Again, operational factors will determine which, if either, may be best for your facility.

REFERENCE SPECIFICATIONS

Security Wire Cages

Wire specification–no. 10 W. & M. gauge 2" times' 1" woven mesh
Post specifications: 2" times 2" square, 14 gauge
Doors: no. 10 W. & M. gauge wire, 2" times 1" woven mesh
    Frame 1 ¼" times 1 ¼" times 3/16" angle
    Integral cylinder lock or external padlock lugs
Maximum height: typically 30'

5R Vault

Concrete: 12" thickness of 3,000 psi test
Reinforcing: Two curtains of no. 5 rebar on 4" centers
Doors: Typically 3 ½" with construction of UL-approved drill and torch resistant material

TL 15 Rating

Minimum weight: 750 lb. Empty
Body: 1" steel with 50,000 psi tensile strength
Door: 1 ½" thickness with strength equal to or greater than body

Alarm systems can range from a single contact on the storage-facility door with an audible device, to elaborate high-security systems with remote annunciation. Alarm companies can be helpful in this area, explaining various system options to you, once they look at your requirements. For many security "cages," the new shock sensors now available for fence applications will work extremely well with a minimum amount of maintenance. In any case, the keys to ensuring the proper functioning of whatever system you may select are (1) regular testing, and (2) proper physical maintenance of the storage facility itself. For example, fence fabric must be kept tight, and doors must be lubricated and aligned.

If security officers are used to deter unauthorized entry, they can also be used to verify items entering or leaving the secured area. If these individuals are from a private contractor, they should normally not be given custody of the area, serving only to verify counts, etc. If your company's own security officers are used, watch for any compromising interaction with other employee groups. In either instance, do not forget to keep an eye on the guard's performance. Assuming that guards are infallible may cause you great embarrassment.

Selection of personnel who will have access to and control over your storage area is as critical as the design and construction of your facility itself. The individuals whom you will entrust with this responsibility should be carefully screened and made fully aware of the nature of their responsibilities. This will then allow you to hold them accountable for the proper control of your secured items, providing that reasonable procedures are established and proper documentation required.

Once your personnel are selected, systems of verification must be developed to ensure that they are properly performing their duties. Again, do not assume anything. Appropriate verification methods include random spot-checks of secured items, unannounced complete inventories, and careful review of any reported discrepancies in records. Additionally, ensuring that an employee takes at least one continuous week of vacation, while a trusted relief person takes his place, may be helpful in discovering any long-term ongoing problems.

DOCUMENTATION

One of the most important aspects of securing high-value freight is proper documentation. Without proper documentation all the physical security precautions and security systems in the world will prove of little value. If you cannot establish what you should have, how can you determine if something is missing? Although this may sound overly simplistic, it is often found to be a key deficiency in facilities where large losses are occurring.

Proper documentation must begin at the point where freight enters your custody. One of your primary initial concerns is to make sure that you have control over your documentation. This can be done through duplicate copies, immediate entries to separate logs, etc. Your key objective is to ensure that all your documentation is never in the same place at the same time, a situation that would easily permit someone to remove first the documentation and then the freight, probably without detection.

Once control of your documentation is established, a verification of quantities received is made, control in your security storage area begins. This again requires proper documentation. Logs of items entering and leaving the secured area must be carefully kept. These logs should be sequentially numbered pages, which are removed as they are completed, and kept secured to afford them adequate protection. Additionally, other controlling documents, such as freight bills and stock cards, should reflect that the items are stored in the secured area, and control of these documents should be assigned to a different individual from the one controlling the security-area log.

In the interest of clarity and operational ease, it is often noted that exact descriptions on controlling documents may be a detriment to your security efforts. Wordings on such documents as drugs, jewelry, watches, etc., will often be all the temptation that a potential thief needs to help himself. If possible, a freight classification number, inventory number, etc., should be used as the identifier. This will at least not identify your merchandise to the casual observer or handler.

The same concept applies to packaging and shipper consignee designations. Many manufacturers of theft-susceptible items tend to have logos, descriptive information, or even a picture of an item on their cartons. This calls unneeded attention to the merchandise. Although it may be helpful for advertising purposes, if the merchandise is theft-susceptible, your advertising may be viewed appreciatively by the wrong audience. Company names can often be abbreviated, or a warehouse or store identifier number can be substituted on documents to further conceal the identity of the merchandise.

ACCOUNTABILITY

Anyone with a secured storage area needs to have one individual designated to have overall accountability for that area. Although other individuals may have access to the area and be involved in its operation, one person should be charged with ensuring that proper procedures are being followed and that proper inventory control is maintained. Due to the nature of his responsibilities, this person should normally be at management level, and have the authority to effect changes that may be necessary for the maintenance of proper security. This security-area "manager" may delegate the responsibilities for inventory verification, procedure review, etc., to others, but he himself should be held accountable for the end results. Below him an appropriate system of checks and balances may be created for the routine operation of the area. Exact procedures will vary depending on the facility involved, but the key to proper security is ensuring that no one person has complete control over both the merchandise and all records pertaining to that merchandise.

TRACKABILITY

Through your documentation and procedures it should be possible to determine the exact status of an item in question. Some companies have greatly enhanced their abilities in this area through the use of computers. Others must still rely on the more conventional paper-file methods. In either case, the objective is still the same—to be able to quickly obtain information on the status of an item or shipment. Your ability to do this is another key to loss prevention. If it is well known that you have the ability to detect a loss within a short time of its occurrence, there will be less opportunity for someone to remove items undetected.

Through your records, it should also be possible to determine the routing or handling of any item. Your logs should indicate who checked the merchandise in and who was involved in its handling, storage, transportation, and delivery. Accurate maintenance of such records is a deterrent to theft by those handling the freight, for they know they would be looked at in case of a loss. Again, to reemphasize, this documentation is of value only if its integrity is maintained. Develop necessary procedures and verification methods to ensure that your documentation is not lost or compromised.

TRANSPORTATION

Transportation of high-value commodities may be by armored car, company vehicle, courier, the mail, common carrier, or any other mode of transportation depending on the circumstances. The transportation mode, like all other aspects of securing high-value commodities, must be selected based upon the risk involved, and other factors unique to your operation. Your ultimate goal in selecting a means of transportation is to ensure that your "risk" is properly managed within the operational constraints of your organization.

Armored-car companies provide services throughout the United States and in most other countries. They normally operate on a regularly scheduled or by-appointment pick-up basis, and they specialize in commodities of high value and small size, including currency, stocks, bonds, precious stones, and precious metals. In dealing with an armored-car company, a number of concerns should be addressed. First, the liability of the armored-car company should be understood. Typically, an armored-car company will assume liability for your shipment from the time at which it is placed in its care and custody at the shipping location, until the time it surrenders custody of the shipment to a responsible party at the destination. This, along with any dollar limits of liability, should be understood before shipments are actually made. Often these are defined in the armored-car company's contract, but even then they should be carefully reviewed to ensure that they meet your needs. Forms for shipping goods via armored car are normally supplied by the armored-car company. But you may also wish to use your documentation in conjunction with these. Documentation normally requires standard shipping information such as name and address of the shipper, name and address of the consignee (receiver), and a description of the items shipped and their value. The forms also contain appropriate areas for signatures to document pickup and delivery of the shipment.

Often items shipped by armored car are placed in sealed or locked bags or containers. If seals are used, they should be properly selected and applied as described in Part Six of this book. Seals must also be carefully inspected by the receiver, to ensure their integrity. Improperly applied seals can permit the easy removal of items from a bag or container without detection. If this occurs, establishing responsibility for the loss can be most difficult.

Most armored-car companies will provide users with signature specimens, or signature photograph sheets of their employees for verification purposes. If you receive these lists, be sure to destroy the previous list each time a new one is received. Outdated lists may reflect individuals who are no longer authorized representatives of the armored-car company, individuals with whom you should not have any further transactions. Some armored-car companies will also provide you with a list of the license numbers and descriptions of the vehicles they use for further verification of their authenticity.

Just because an individual is an employee of an armored-car company does not mean he should be given free access to your secured areas. Most organizations designate an area of their facility for transacting business with armored-car personnel. A common practice is to ensure that if that location is in one of your secured areas, a representative of your company is present with the armored-car personnel at all times.

Rates for armored-car services vary substantially throughout the country, and even in a single geographic location. Once you decide that the cost of armored-car transportation is justifiable for your organization, and such service would best suit your needs, compare the costs, service, and reputation of the various companies in your area. Then you can make a factual decision on which company would best suit your needs.

Another means of transportation for theft-susceptible items is the common carrier or courier companies. These companies account for the movement of a tremendous amount of freight in both intrastate and interstate commerce, and the nature and value of some of the items being transported by that plane, train, or truck that we are so used to seeing may amaze you. Can these "common" modes of shipping be safe for high-value or theft-susceptible items? The answer is often yes, if certain precautions are followed. One of the most important considerations in shipping via a common carrier is your "advertising." Although this was covered earlier in the chapter, some reemphasis at this time seems appropriate, because your freight will be handled and transported in a relatively unsecured environment. By "advertising" I mean the way in which cartons are marked, merchandise is described on freight bills, etc.

If you are shipping high-value or theft-susceptible items via common carrier, and your firm insists on prominently labeling the outside of cartons in such a manner that it will be fairly obvious what they contain, you are asking for problems. A review of your claim history may provide you with sufficient ammunition to do battle with your marketing department over carton labeling. Sometimes they lose sight of the fact that losses, due to theft, also have an effect on the profit-and-loss statement's bottom line.

If you need more ammunition, you may wish to contact the security or claim-prevention personnel for the major shippers that you deal with. They will often be able to provide statistical support and other helpful ideas to you. They share the common goal of theft prevention with you, and usually will be anxious to help. Other "advertising" areas that should be monitored include the descriptions on freight bills, and shipper consignee names. Instead of listing an item on a freight bill as drugs, guns, jewelry, etc., classification numbers would be much more effective in reducing the chances that the item may be stolen. The same applies to shipper consignee names. Scenic River Firearms or Johnson Pharmaceutical Company, graphically descriptive names, may be best abbreviated on a freight bill as SRF Inc. or JPC.

As a general rule, the less shipping-related "advertising" you do, the better the chances that your high-value or theft-susceptible item will "live" to reach its destination. When you ship by a common carrier, also remember to check on any "released-valuation" rules that may pertain to your product before you make your final decision on the economic feasibility of shipping by common carrier. As an example: If your product has a released valuation of 50 cents a pound but the 10-pound item is worth $500, the economics of shipping in this manner may prove to be less than cost justifiable if you suffer many losses.

Another mode of transportation utilized with high-value theft-susceptible items is company courier. This mode of transportation is used where it is felt necessary to retain possession of an item internally to ensure its safe movement. Items often shipped in this manner include gems, precious metals, stocks and bonds, cash, and radioactive materials.

The specific means of transportation and procedures to be used when moving items via company courier are highly dependent upon the commodity and the specific needs of the courier. The shipment may also be escorted or unescorted. In any case, adequate and well-defined procedures are needed to ensure a successful operation. With many commodities a single courier is used without an escort. An example of this would be the pharmaceutical wholesaler that uses delivery vans to move commodities to retail drug stores. This courier should operate on a predetermined route and time schedule known only to necessary in-house personnel. The route should also vary on a day-to-day basis. In conjunction with this, a "call in" system should be adopted utilizing either two-way radio or telephone. The courier then reports in at prearranged times. In conjunction with the prearranged routing and schedule, this information will permit company personnel and police to at least know in what area to begin a search, should a problem occur or the courier not call in. This individual should also be provided with instructions pertaining to the actual delivery at the customer's premises. These instructions should include (1) keeping a watch for suspicious persons and/or vehicles and what to do if they are spotted; (2) appropriate locations for parking, the more visible the better normally; (3) securing of the vehicle while it is unattended, and use of any vehicle alarm systems; (4) the most secure means to physically carry a package while walking–tight and close to the body; (5) how to deal with an attempted holdup; and (6) other procedures as appropriate to the safe delivery of the commodity.

Other movements may include escort personnel in varying numbers. An example may be a transporter of hazardous waste. Such a movement will often involve a three- or four-vehicle convoy. This group will include a lead vehicle, the transporting vehicle, a tail car, and possibly a backup power unit (transporting vehicle). All units are typically in radio communications, and security personnel are in each vehicle in appropriate numbers. The lead vehicle precedes the transport vehicle to check road conditions and to watch for any possible problems. The tail car monitors activity behind the transport vehicle, often maintaining separation between the transport vehicle and unidentified following vehicles. Again, each unit and involved individual must have specific instructions on dealing with situations and emergencies that may arise, from routine refueling stops to attempted sabotage.

The use of company personnel must be carefully evaluated to determine if it is the most appropriate means of transportation for your commodity. Included in your evaluation must be such factors as operating costs and personnel-liability costs. If the commodity is particularly vulnerable to theft, consideration should also be given to the appropriateness of possibly placing company personnel in a position with an inherent element of danger.

Another means of transportation that is often ignored, which may be suitable for small items, is the mail service. Registered and insured mail are used by many organizations to move certain theft-susceptible items. To determine the viability of using the mail service, you should check with your local postal officials on rates and shipping restrictions, and you should ensure that your insurance coverage does not prohibit such movements.

DEALING WITH A LOSS

One of the areas of concern that is easy to overlook, usually out of hope that it will never happen, is what to do if a loss occurs. Although the action you should take will be dependent upon your industry and the commodity involved, there are a number of general items that should be considered in any case. These include planning, notification, and publicity.

Planning what to do in case of a loss is a critical function for anyone dealing with theft-susceptible commodities. Your plan should be explicit in defining what must be done, and who will do it if a loss occurs. Without such a plan, it is easy to overlook many important steps that should be taken. A typical plan would look something like this in outline form:

Item		*Responsibility*
1.	Notify police	Security
2.	Notify management	Security
3.	Interview involved persons	Security
4.	Notify insurance carrier	Financial Officer
5.	Determine amount of loss	Accounting
6.	Insure serial numbers entered in NCIC	Security
7.	Advise customer and arrange replacement	Marketing
8.	Write loss report	Security
9.	Follow-up	Security

This would then be expanded to include the necessary specifics, such as, which law-enforcement agencies should be advised, which management personnel need to be notified, and other details pertinent to your operation.

It is fairly obvious that notification is one of the most critical elements in your response procedure. These notifications to the law-enforcement agency that will investigate the loss, and to company management, who will act to minimize the effects of the loss, must be made promptly and accurately. Records should then be maintained indicating the date, time, names, and phone numbers of those notified.

If your shipment was being made interstate, or was of a specially regulated commodity such as guns, drugs, or stocks, federal law-enforcement officials should be notified in addition to local agencies. The FBI investigates thefts from interstate shipments. The Bureau of Alcohol, Tobacco, and Firearms (ATF) investigates major firearms thefts. The Drug Enforcement Administration (DEA) is interested in major drug thefts. Know in advance if any of these or other agencies are in a position to assist with the investigation of your loss.

If your commodity has serial numbers, be sure that these numbers are provided to the investigating agency for entry in the FBI's National Crime Information Center (NCIC) computer that stores information on stolen items. Most local law-enforcement agencies have access to this data, making it important to have your information entered in case a policeman somewhere should locate a part of your shipment and make an inquiry through the system.

Other possible notifications would be to your insurance carrier and to industry sources. Many insurance policies require notification within a specified number of hours following discovery of a loss. This notification should also be documented. Some specialized industries, where the commodity taken would be difficult to dispose of outside industry sources, may also have a cooperative network for sharing loss information. One of the most successful of these has involved a major typewriter manufacturer whose service facility checks units brought in for repair to determine if they are stolen. Your industry may have a similar program.

Once necessary notifications have been made, the investigation and documentation of the loss become a prime concern. Your established handling procedures should permit you to establish the routing and handling of the shipment with minimal effort. With this information in hand, it will then be appropriate to interview those involved in dealing with the shipment, obtaining written statements whenever possible. This effort should be coordinated with law-enforcement officials to better ensure that the common goals of identifying those involved in the loss and recovering the merchandise are achieved. This may mean more paperwork, but those statements may be a key in establishing what happened and who was involved.

Documentation should also be prepared to support your claims on the nature and amount of the loss. This information will be needed for any criminal proceedings and most likely will be requested by your insurance carrier.

If a solution to the loss does not develop once the basics have been uncovered, follow-up and continued liaison with law enforcement are the usual course of action. Review what has been done to ensure that nothing has been overlooked. Keep in touch with the investigators to ensure that their efforts are being supported as best you can. And have patience, because sometimes investigations can take a great deal of time. The end results may well be positive.

As with all of your high-value theft-susceptible commodity security program, what you do in case of a loss should be carefully planned in advance. This will lead to a minimization of the frustration in dealing with such an occurrence and a maximization of your efforts to recover the merchandise. And if a theft happens, be sure to review and update your plan, based on your experience in its use. Fine tuning is an ongoing, never-ending process.

SUMMARY

This chapter has attempted to give the individual with a minimal security background an overall view and plan for dealing with the protection of high-value or theft-susceptible commodities. Other chapters of this book deal with a number of facets of such a program in much greater detail, and should be utilized in detailing your security program. In any case, remember that the key to a good program is adequate planning supported by appropriate documentation, and a great deal of common sense. And if at first you do not succeed, try, try, again!

GROUND TRANSPORTATION AND TRACKING

At the heart of all transportation, regardless of the transit mode, is the cargo in transit. The process of transporting cargo from the point of origin to the point of delivery is a complex one that traverses multiple transit modes, intermediate processes, facility owners, and national jurisdictions. Shipments can be subject to various misfortunes such as improper handling, improper routing, theft, and damage in transit.

The international environment is more complex and integrated than at any other time in history. Concurrent with today's global economy, technological advances such as electronic data interchange and just-in-time delivery practices have added to the intensity of commerce and trade. Traditional safeguards that accompany the shipment of cargo have advanced to accommodate the rapid pace of today's commerce. While the process of cargo transportation has been improved in the interests of expediting its transportation and documentation aspects, the security aspects of cargo transport have not. Given the nature of today's transportation environment, new opportunities for exploitation exist that could allow for a wide array of cargo loss. As a result, the need to apply technology to mitigate the amount of cargo loss or theft is required. The following areas represent several approaches in using information technology to reduce these risks.

Smart Cart Technology

Today, most of us log onto our computers with a log-in and password sequence. In the very near future, we will use smart card technology to access our desktop, laptop, or any other computer needed to conduct business. The smart card has begun to provide the appropriate security levels that will be needed in the new millennium. Digital signatures and digital certificates will be stored on these cards and will afford the end-user the ability to use any computer, via the Internet or an intranet, to access the desired secured information. Many of the security issues dealing with access to a company's electronic data interchange (EDI) network may be minimized by implementing smart card technology. The resultant conclusion would be to reduce the number of unauthorized "insiders" that presently have access to bills of lading and other critical information that is in electronic form. Applications of smart card technology promise significant benefits for identification and access control.

The business case for smart card use is built around the core function of authenticating user identification. All terminals and companies share the common function of verifying the identity of those who use their facilities, networks, and information services, and verify the authority of those individuals to exercise certain tasks and privileges. Card-based systems have proven especially effective in performing the general function of verifying an individual's identity and authority to act. The portability of smart cards means that the cards can be used at the user's workplace, off-site facilities, or remote locations to provide strong user verification and standard access to transportation services. Enhanced data storage capability of integrated circuit chips, and identification data may be stored in several forms on smart cards. This data could not only include an individual's biometric template, but also the cargo manifest or bill of lading information. Much of the information that is presently stored in a record located at the file server or client level will now be stored digitally on the smart card. The one key feature that adds so much functionality to the security aspect of business is the public key certificate that binds the cardholder to a unique name and digital signature. This capability allows an individual to log into a computer at any location, which is connected via the Internet or intranet, and conduct secure transactions. This is also true for e-mail as well. Whenever a user logs into a computer using his or her smart card with the public key certificate, all the rights and privileges of the individual are now resident in that terminal.

Smart cards used for identification and access control usually fall into six major categories: (1) the level of security, (2) data elements/database, (3) hardware/software, (4) systems integration, (5) application of biometrics, and (6) additional card applications. In addition, the card features such as its standard appearance, distinctive terminal, port or facility identity, interoperability, and multiple functionality need to be addressed.

Whether we are controlling computer access, network access, facility access, or making a telephone call, smart card technology will be the way in which that access is authenticated and will be the way we conduct our everyday personal, business, and banking transactions.

Automated Gatehouses and Access Control Systems

Automated gatehouses use information systems to match cargo, paperwork, trucks, and drivers at exit points, and are designed to reduce the illegal movement of goods out of facilities. The Seagirt Marine Terminal at the Port of Baltimore uses a computerized gate complex. This computer system serves as the nerve center for terminal. Seagirt's automated system consolidates the steps necessary to generate the trailer interchange report (TIR). When trucks enter the terminal, an electronic sign-bridge over 13 of the 14 inbound lanes directs the drivers to the appropriate lane, where a remote intercom system allows them to quickly exchange information with clerks in the gatehouse. For export traffic, trucks are directed to one of four lanes with built-in scales. While the driver supplies the clerk with the necessary information, the container's weight is automatically put into the clerk's computer. The driver then pulls under the gatehouse canopy, where a terminal mechanic inspects the container and chassis for any damage. Should any be found, the Seagirt computer prohibits the clerk from reassigning the equipment until it has been repaired. The driver then receives the TIR and is released into the terminal. The process for import loads is essentially the same. The clerk verifies the driver's information and that the container has received all the appropriate releases. The clerk then issues the TIR, informing the driver where the container is stored. After picking up the container, the driver goes to one of the five outbound lanes, where a security check and final inspection are performed.

Advanced Automatic Equipment Identification (AEI) Technology

Electronic data interchange (EDI) systems facilitate the movement of goods through many ports. These systems can provide the ports with arrival notices from ocean carriers, delivery orders from customs house brokers, cargo status replies from marine terminals, and electronic bookings from freight forwarders to ocean carriers.

Automated equipment identification (AEI) technology uses proximity tags both active (battery inside) and passive, or barcodes and barcode scanners with radios to transfer real-time data on shipping container and/or vehicle locations. Many carriers use AEI technology for container identification, chassis management, automatic container/trailer weighing, gate access control, and intermodal movement operations. Benefits include accurate data entry, faster location of containers in terminals, and reduced work force requirements.

A Matson terminal located in Honolulu has implemented AEI. Small durable transponder tags are attached to every truck cab in use by the terminal. As a vehicle approaches the front gate, the tag is queried by a transceiver, co-located at the entrance. The tag responds with time and identification data, which automatically enters into Matson's database. Next, the vehicle rides over a scale and its weight along with shipment information is then computer matched to the established record. Upon leaving the terminal, the driver approaches the exit gate, again being queried by a transceiver, and a hard copy of the transaction is printed along with the turn time in the terminal. At that point, the driver is allowed to exit.

This approach to automation translates into a real return on investment (ROI) when compared with the time required to process a gate pass. The manual system would take about one minute to process, while the AEI system completes the same process in fifteen seconds. Approximately 250,000 transactions a year occurred at this terminal and resulted in over 3,000 work-hours saved. Not only does this provide for an accurate inventory count of trucks that have entered the terminal, but also any truck that remains too long within the terminal is flagged by the computer. This warning affords the superintendent an opportunity to investigate the delay. This inherent feature provides additional security to the other containers located within this terminal. With over 200,000 transactions in its first year, the system had no misreads.

With an AEI system, the terminal operator provides a service to its customers. The technological advances of AEI provide faster turn-times for the trucks, increase inventory control, and enhance container security—a benefit to everyone, except the perpetrator!

Tracking Goods

All of these technologies help to pinpoint the location of goods at each step in their movement from origin to destination. To find the location where subversive activity occurs is paramount in order to focus the required security resources appropriately. The primary concern for security management is internal theft.

As a result, one overnight delivery company has integrated a digitized video system with its automated scanning system. The standard processing procedure is to scan each package several times from the moment it is picked up until the time it is delivered. The company then keeps track of exactly where the package is at all times through its package tracking system. The most common place for internal theft (according to this company) is at the destination station, where the drivers remove the packages from the conveyor belt and place them in their trucks for delivery.

Theft of these packages is not difficult to execute. An unscrupulous driver could simply take a package and place it on the truck, even if it is not on the required route. If the driver does not scan the package, signaling to the package tracking system that it is on the delivery truck, the package is lost without record. Since there can be 80 trucks at the destination station, it is difficult for management to figure out on which truck the package was placed.

In the past, video cameras were placed above the conveyor belt to record the packages moving along the belt. Unfortunately, searching through the video recording to find the exact moment the package was in view of the camera was difficult and impractical. The quality of video suffered after multiple recordings, and a VCR maintenance program was necessary to ensure the continued quality of the recorded video. Furthermore, the manual changing of VHS tapes was unreliable and video recordings were often lost. An enormous amount of personnel and time were required to bait and track specific packages and to conduct truck audits in search of the missing packages. Thus, a different and innovative solution was needed.

What one digital video company has done is to associate the digitized video of the packages moving along the conveyor belt to the package tracking number. Video retrieval from a PC-based system requires only the tracking number.

It is possible to link a package scanned at the beginning of the conveyor belt with the video of the package moving along the belt. This enables quick retrieval of package video and reduces the time it takes to investigate. Furthermore, this process completely automates the recording and long-term storage; therefore, there is no need for human intervention to record and store video. Video is recorded straight to hard disk, then transferred to an advanced intelligent tape (AIT) and automatically stored in the tape library, which is essentially a tape jukebox. The internal robotics change the tapes as soon as they are full, without human direction. The delivery company is now afforded video for the past 45 days. No tape changes or human intervention is required.

From the review station, an operator simply enters the tracking number and all the video for that package is automatically retrieved. One keystroke combination brings the next camera in the sequence up for review. The package can be viewed as it travels down the conveyor belt and can be seen as it is removed and placed in the delivery truck.

Because of the placement of the cameras and the ease with which the package can be viewed as it moves down the conveyor belt, anyone who takes the package from the conveyor belt can be readily identified, as well as the vehicle in which the package is placed.

Once a suspicious handling of a package occurs, the video is exported. The video export feature allows the video to be saved indefinitely on VHS tape or recorded on a CD-ROM. Delivery companies are now able to access strong evidence when a package is inappropriately handled in order to prosecute the thief.

Part Six

SEALS

SEALS

CARGO SEALS

Cargo seals are theoretically designed to serve two purposes—to discourage theft and to provide positive indication that a theft or theft attempt has been made. Today we have two types of cargo seals. One is the simple strap seal that basically depends on its tamperproof design or its frangibility; the other is the high-security seal. The simple strap seal is *not* a lock, but rather a numbered control from point A to point B. A simple seal also will not stop anyone who is determined to get into the unit physically, but proper control and use of the seal will deter someone who is thinking of taking something without anyone's knowledge. A seal's effectiveness in discouraging theft is directly related to the determination of the thief. A good seal should provide clear evidence if it has been tampered with or illegitimately opened. If it can be opened and closed without leaving clear evidence that this has been done, one might as well use a bow of string or pretty ribbon. A seal also must be strong enough to prevent accidental breakage during normal use, and its design must be complex enough to make unauthorized manufacture of replacement impossible.

A seal must be tamperproof to the point that its forcible removal would result in the visible destruction of its essential parts, and therefore it must be embossed or marked with letters and numbered for document control. Actually, seals are probably the least expensive tool today in loss prevention, and yet they are the most ignored or overlooked. A four-cent, simple seal, properly controlled and documented, can usually indicate the point of loss from point A to point B or from the time of closing the trailer or container to the opening of the unit at the ultimate destination. In the case of over-the-road relays or in container shipments from water ports to trailer and container exchange areas, warehouses, terminals, or consignees facilities, the seal integrity, if checked, will absolutely show the point of seal violation. If a unit goes from points A to B to C to D to E, and seal control shows the seal intact at C but not at D, it is quite logical to assume that the violation of the seal occurred between C and D. Therefore, the investigation will be contained in the area between C and D, negating a costly investigation all the way back to point A. If one has tight seal control and has checked the seal at all points, look at what one gets from a seal costing a few pennies as opposed to costly investigation at all points.

An assistant vice president of a major railroad recently revealed to me that by enforcing good seal control according to written procedures, and by *demanding* that the railroad personnel check seal numbers against the paperwork at pertinent points, that railroad was able to deny and absolve itself of claims amounting to $3 million in one year! Yet, it is a fact and has been proven many times that seal control is extremely lax. One sees seals lying around the dispatcher's table or desk, hanging on a nail on the shipping platform, or hung on the handle of a trailer in the process of being loaded over a 16-hour period. Why? We certainly do not leave our company money lying around like that; but then again, one might say that indeed we do, for an uncontrolled seal subject to tampering or switching prior to use, *is the company's money.*

Today there are several reputable firms manufacturing seals. Some are metal, some are plastic, some are prettier than others, but it all comes down to the fact that most of them will do the job *if* proper use and control are maintained.

HIGH-SECURITY SEALS

The conventional seal is no longer respected in most areas of the world. The flagrant abuse of seals during the last several years led to the development of a heavy-duty, self-locking, initialed, and serially numbered cable or heavy steel seal. This unit is highly recommended for all shipments traveling through known problem areas.

High-security seals or cable seal locks provide physical security as well as undisputed seal integrity. This dual-security device should be carefully considered for certain phases of transportation operations. There are several security seals on the market today. They vary in price and strength, and the ability to withstand tampering, and they are made in a variety of designs. I guess it all depends on how much you want to spend on security.

Seals reduce theft claims anywhere from 35 to 95 percent, according to studies made in the United States. Yet in far too many instances they are overlooked by security officials and management as too much trouble or too expensive; but high-security seals are designed to preclude the efforts of amateur thieves and vandals, who account for the majority of theft claims reported by carriers annually.

The professional thief will hit you eventually, regardless of security measures taken, but most theft and damage claims are caused by the amateurs who prey on unsecured shipments. When U.S. Customs first tested the initial cable seals, 495 trailers containing over $36 million in freight (an average of $69,000 per trailer) were shipped throughout the United States under Customs control. After the test, when 495 trailers and containers containing over 60 million in freight (an average of over 121,000 per trailer/container) were shipped throughout the United States under U.S. Customs Control, only $264 worth of freight was missing! There were a lot of chisel, hacksaw, and hammer marks on a lot of the seals, but only one was missing! The test being conclusive, U.S. Customs approved the first cable seal.

Some trucking companies feel that high tech, high-security seals require too much of a capital investment, but how can one argue the point when we hear about a major transportation company in the United States that reduced its liquor and sugar losses to zero, and another that reduced its overall claims by 60 percent by using a high-security seal? In a U.S. Department of Transportation study involving railroad cars of tires it was determined that for every dollar spent on high-security seals, $14 were saved because shortages were proved as short-shipped. It is imperative to implement a process that can respond to the electronic notification that a high tech seal has been compromised en route.

One can and will say that a high-security seal flags a hot load. While this is possibly true, it also "buys you time." The time it requires to take off that high-security seal should now be considered as one of the reasons why you bought it—it is buying you valuable time, because the thief needs tools and time. If the "pro" wants the load badly enough, he will take the entire rig, but seals slow down the semipro or amateur into a period where he might be observed, and he just cannot take that chance. An extra few seconds to remove a seal is the time frame in which he may be caught.

Have you lost some freight en route from here to there? What was the number of the seal that was put on the truck? Probably 40 percent of trucking executives and managers do not know, just by looking at the backs of the trailers traveling the highways without seals. Another 25 percent of them do not know because the seal number was not recorded in a permanent book, and another percentage of them are not really sure because it seems that the dockman who loaded the truck put the seal on instead of his supervisor.

Yes, the same man who possibly shorted the load by five stereos was handed the seal to do management's job, because the supervisor was "too busy"! Does that sound familiar? That is the way it often happens today. Mr. President/Manager: Take a close look at your operation—or are you one of the few that can stand up and be counted as doing it right? Just imagine, do you or your family leave the house without locking the doors? Of course not, yet you may let a trailer or container out of the yard, in many cases without a seal. Most drivers like a sealed trailer because it shows that the load that they took out from the terminal is still intact, as far as they are concerned. In fact, if he is smart, the driver will have the consignee break the seal or have someone check the seal before he himself breaks it. I know that many drivers "pretend" to break the seal in front of the consignee, who is fifty feet away, up on the dock, so that he can swing the doors open before backing in—but here again the seal is a mental deterrent, otherwise the driver would not go though the charade in the first place. This alone means that the seal system has some effect.

As I have said before, it is imperative that information regarding each seal be available throughout the life of that seal. In other words, the procedure so vitally important is to control, document, and trace the seal from the point of receipt from the manufacturer to its ultimate destruction. Here are a few time-tested tips that may assist you in seal procedures and control.

Ordering and Storage

One person only should be charged with the responsibility of buying protective seals from the manufacturer. The manufacturer should be instructed to ship the seals to one person's attention only at the home office or another designated place. The company's name or initials should be embossed onto each seal. It is also recommended to code each terminal by letter, numeral, or color.

Distribution and Outbound Recording

One person for each shift should be charged with the responsibility of dispensing seals. It is *not* recommended that drivers be given seals to secure the trailers they drive out of the terminal yard. Effective control is possible only with accurate records of pertinent data. The information suggested for recording in the outbound mode is on the manifest or load sheet, which will be retained for some time to compare numbers from origin to destination in cases of shortages from the unit. Information to record includes the following:

1. time seal is applied
2. load destination
3. date number of the trailer
4. name of person applying the seal

Drivers making pickups involving shipper's load and count can be furnished with a seal to be applied at the time of closing of the doors by a shipper's representative. The driver will show on the bill of lading or shipping order the seal number applied at the shipper's dock.

Application

Unless seals are properly applied, control is lost over your sealing system. The procedures to follow in closing are simple, safe, and fast.

1. Seal all doors—not just the rear ones.
2. Run the seal through the hasps only once. Seals wrapped around several times become illegible.
3. To ensure positive closure, tug down on the seal with sufficient pull to make sure that it is locked.

Gate Procedures

If gate operations are part of your organization, the security officer should record the following information, in addition to pulling on the seal.

1. Tractor number.
2. Trailer or container number.

3. Seal number, color, and coding.
4. Driver's name.
5. Date and time.

On high-security shipments it is also suggested that the security officer apply a second seal. Some low-valued shipments could also be double-sealed to prevent an obvious pattern from developing.

Broken Seals

At times, it may be necessary to break a seal between terminals or between terminal and final destination. An occurrence of this nature must be reported and the following information recorded.

1. Name of the person breaking seal.
2. Reason for breaking the seal.
3. Time and date that the seal was broken.
4. Serial number of the broken seal and the serial number of the replacement.

Whoever has been instructed to break the seal should do so in the presence of a witness. If witnesses are present, their names must also be included in the broken-seal report. This broken seal report must be filed with the terminal. This procedure must be followed no matter how far the driver may be from the point of origin.

Seal Removal and Inbound Recording

Many shipments are received and assumed intact if the seal is seen swinging from the door latch. Nothing could be further from the truth. There is only one way to make certain that seals have not been violated, and that is by a physical check. The recommended procedures for seal removal are as follows.

1. Authorized personnel only should remove seals.
2. The name, serial number, and all coding information appearing on the seal must be checked against the corresponding shipping papers and recorded on inbound documents or manifests.

3. Prior to breaking the seal, pull down on it, making sure that it has not been shortened or faked.
4. Any discrepancies in information on the seal or on the shipping documents should be reported to persons assigned to accept such statements.

Should a shipment be received with a broken seal or a seal violated in any manner, reports must be turned in as previously described. Any evidence of theft should be reported to the security department and an investigation begun immediately, irrespective of the time of day or night.

Other Procedures

There are other steps that can be taken to beef up your seal-control program. Some of the suggestions can be part of your security system, whereas others are optional.

1. The door hinges and locking handles must be such that removal cannot be made without violating the seal.
2. Sealed loads discovered to be tampered with while in transit should be immediately reported to the nearest terminal location or location assigned to receive such reports. High-value loads in the yard should be inventoried and controlled.

AUDIT PROCEDURES

I am sure that most managers do most of this with their eyes closed. Yes, maybe some really do! When is the last time you checked to see that the seal procedure is being followed to the letter? How about a spot-check at the gate? Have you ever seen any unused seals lying on the dock floor? Are the seal numbers going on documents, or do some terminals get blamed for shortages because no seal number is available from origin? I guess we could examine the system at any time and find human error! So, in essence, control your seals at all times—treat seals as money. Maintain good seal records. Check your seal numbers against documents. Come down hard on anyone not complying with seal procedures. *Strict* seal control and proper use of seals will pay off. With regard to the railroad that saved the $3 million dollars—it was because they tried, and it worked!

USE OF SEALS

1. Inspect seals whenever a sealed containerized shipment enters or leaves a facility. If the seals are not intact or there is evidence of tampering or the seal numbers are incorrect, notify security and/or management personnel and tally the cargo.
2. Seal unsealed containerized shipments at the point of entry to the facility and note the seal number on the shipping documents. Seal all containerized shipments leaving the facility and note the seal number on the shipping documents. Inspect the contents of containers received in an unsealed condition.
3. Release seals to as few persons as possible. Require all persons handling seals to maintain a strict control of the seals assigned and to store them in a secure place.
4. Maintain a seal distribution log, which indicates to whom seals have been released.
5. Where possible, secure containers by butting or "marrying" their door ends against each other. However, do not butt them against a perimeter fence or building wall if that will compromise the protection provided by the fence or wall. In stacking containers, place those containing high value merchandise on top. Finally, containers or trailers can be backed up tightly against a loading bay door to protect from thieves who do not have access to the warehouse.
6. Locate high-value merchandise in mounted containers or trailers in a special security holding area where it can be observed by management and/or security personnel.
7. When containers are mounted on chassis, secure fifth-wheel by a pin-lock which meets the minimum standards for locks and is constructed to withstand normal abuse from equipment. Hold designated management and/or security personnel responsible for storage and control of pin-locks.
8. Restrict access to special security holding areas and permit the release of containers or trailer from such areas only in the presence of management representatives and/or security personnel.
9. Select seals for use that are consistant with the level of integrity sought associated with their use and application.

10. When inspecting a seal always examine the hasp, hinges, rivets and doors.
11. Destroy all used seals and seal parts. Never leave them on loading docks.
12. NEVER believe there is such a thing as a tamper proof seal.

Part Seven

INVESTIGATIONS

INVESTIGATIONS

LOGISTICS INVESTIGATIONS

The marketplace of today is truly global. Whereas in the recent past, investigators were not necessarily required to possess detailed knowledge of manufacturing and distribution. Nowadays companies often perform some integral pieces of their business outside the supply channel. This business condition mandates that the investigator understand the processes of manufacturing and distribution, so as to provide the most accurate detailed accounts relative to losses. As losses occur, all parties who are involved will look for a focal point to link critical pieces of investigative data. Corporate security investigators are well suited to serve in this capacity. Investigators must be able to provide insight pertaining to manufacturing and distribution locations, as well as have contacts that can provide detailed product information.

Trending Analysis

Obtaining meaningful information which indicates detailed loss data will assist corporate security efforts in stemming the tide of losses. Often, investigations will reveal the need for process improvements in various areas of the company. In many instances, the investigator will discover what may be described as "compartmentalization," where smooth transitions between departments are required for lessening the company's exposure to loss. Many data fields may be scrutinized, however, some of the more important measurements are as follows.

Internal Measurements

What level of inventory variance is acceptable and how is it reported?

Who conducts variance reconciliations where discrepancies exist?

How are shipper/carrier counts conducted?

Who has the authority to re-create bills of lading?

What specific security controls are in place throughout the process?

How are these controls administered and measured?

Internal measurements and controls are designed primarily to keep the shop honest, yet security professionals must temper their efforts with the needs of the business.

Carrier Measurements

Who provides transportation services for your company?

How is their success determined?

Price and service vs. rate of defect or losses

Are measurements conducted company-wide, by product, by shipping lane, by carrier, or some other way?

Does the security investigator have access to data?

Carriers and some internal organizations may present data that make the loss problem appear extremely small. This is achieved by representing the percentage of loss against the total dollars shipped. Now, those who provide statistical data are well aware of this risk, particularly where corporate operating ratios in the high-tech arena may be less than five percent. As security professionals, there is an obligation to demonstrate the return on investment when losses are significantly diminished.

Process Improvements

As a result of trending analysis, the loss fields should become readily apparent and pave the path for process improvements. Where data indicates particular problem areas, the security organizations should move on two fronts. The first is internal departments where analysis indicates the need for process improvement. Interdepartmental or team meetings may be in order to correct shortcomings. Participation in such meetings is time-consuming and usually includes cross-functional training or awareness in an attempt to improve an existing defect. This situation is most prevalent within large organizations because work is necessarily segmented thereby creating segmented processes which do not always create smooth transitions.

The second is business partners. Businesses with which other companies are closely linked may need to work together toward mutually beneficial problem solving. For example, manufacturers and carriers may both be viewed as experts at what they do, however, each tends to point the shortcomings of the other when losses are concerned. The end result of this example is often the termination of the business relationship. This approach not only fails to solve the problem at hand, but the problem will certainly be replicated with multiple players. Cross-functional meetings with the business partner are in order in this scenario also. The teams can meet in one another's element where they can touch the points of contention and seek resolution.

Front-end process improvements will always diminish the potential of loss from systemic defect. This will enable law enforcement and security professionals the latitude to work on cases where theft is the primary cause for loss.

Opportunities for Theft at the Manufacturer

Cargo is most susceptible when it sits idle. From when product is manufactured until the consumer receives it, opportunities for theft are greatly increased. Manufactured inventory is an inventory that is shelved in preparation for shipment, but drawing from the inventory without a customer order makes it vulnerable. Manufacturers of high-tech products may follow build-to-plan production schedules. In either setting, vulnerabilities exist. Let's explore both.

Build-to-Plan

Sales and forecasting of consumer demand dictate how many units will be assembled under this business condition. These types of operations are generally giving way to the build-to-order method, however, they are still in force across many industries. In this setting, large warehouse and assembly operations mass produce items in assembly line fashion. Upon completion, customer orders which are outstanding, or even forecast, become fulfilled for immediate shipment. Any overages are either shipped to other areas of the business which may have an outstanding need, or the inventory is subject to theft. Cargo theft investigators must determine at which point in the supply channel the theft occurred. If the theft occurred prior to shipment, the investigation becomes more complicated, either a large amount of product will sit at the manufacturer awaiting shipment orders, or in the supply chain awaiting time-sensitive delivery.

Build-to-Order

Business conditions have, in large part, evolved to this manner of production. Build-to-order production calls for prompt customer service from the manufacturing environment. The manufacturer must take the customer order, produce the product, and place the order in the supply chain in virtually the same time as the build-to-plan method. Build-to-order calls for prompt infusion of parts and services from sources close to the operation. Almost no finished product sits idle under this plan, however, other opportunities for the thief exist.

Many different groups are needed to enable this process. Often contractors or off-site vendors must supply parts or provide services under short notice, and in bulk. This creates a critical element to theft. When the thief has access to the process, the possibility of theft increases.

Control of process is another important issue. Where time is an essential factor in production, the manufacturer must release some control in order to achieve production goals.

Opportunities for Theft With the Carrier

Most carriers of high-tech cargo do a good job with respect to pick up and delivery. The age-old battle remains regarding freight rates and specialized handling requests. The manufacturer desires the most cost-effective rate by the carrier in bringing product to market and the less-than-truckload (LTL) carrier looks to consolidate the most desirable freight from multiple sources to a single location. Although the process sounds simple, problems are inherent.

At the point of pick up from the manufacturer, a careful timing of trailers, carrier drivers, and loaders must take place to meet production demands. The carrier must provide sufficient trailers in a narrowly timed window to achieve established goals. Due to the timing factor and volume of shipments, product is often not counted or miscounted which establishes the first opportunity for theft. If personnel from either the manufacturer or the carrier realize that an opportunity for theft exists, it will be pursued.

The less that cargo is touched or moved from production until delivery, the greater the success rate will be upon arrival. The bill of lading (BOL) should be used as the appropriate tool in examining a loss. The BOL is examined and agreed to by both the shipper and carrier at the point of pick up. The BOL is also used to tender the cargo to the customer at the delivery point. Barring any noted exception to the BOL, it stands to reason that the shipment should arrive intact.

Wherever the freight stops, it should be examined for loss. Almost all trailers travel with seals or locking devices. Most carriers require inspections of the seal or lock at each stop that is made while in transit. If and where this process breaks down should be the first point of hands-on investigation.

During the course of transit, the cargo will often move through distribution centers which are main activity hubs for the carrier. These hubs are often well protected with fences, lights, and guards. Additionally, records of trailer entry and exit may exist which note the condition of the seal and should verify that the shipment safely arrived to this point. More and more, carriers have electronic records of the shipment as it moves from point to point in the supply chain. This could be an indication of a problem if the shipment does not tally as expected. This would give investigators a good estimate of when the shipment became incomplete and would also provide names of possible witnesses or perpetrators.

End-of-line terminals provide targets to criminals as they are often located in high crime zones of less populated areas. In many cases, these terminals do not possess the security resources that larger distribution centers have. Weekends and late nights become prime opportunities due to reduced staffing during those hours. Substantial thefts are easily achieved at these locations which creates two avenues for law enforcement. One is proactive patrols. Since the terminals have little or no staffing during much of the weekend, suspicious activity is easily detected by proactive patrols. The other avenue involves sting operations. Sting operations in high crime areas may stem the tide of increased criminal activity.

At the customer location, many carriers are required to drop or leave trailers, sometimes up to several days prior to unloading. This creates two basic opportunities for theft. The first opportunity comes into play where the trailer is parked. Adequate fencing, lighting, and security precautions should be employed so as to minimize theft potential. The second opportunity is presented when the shipment is actually unloaded. Where loss is detected, the investigator should give consideration to the entire path that the shipment has traveled. In the event the shipment moves from the manufacturer through the carrier's network, close attention should be given to where and when the seal was removed from the trailer. Additionally, the carrier driver will in many cases assist the consignee or customer in counting the shipment. This provides the investigator with several potential witnesses.

Theft Investigation

There are two primary approaches to investigating high-tech cargo thefts. A major loss occurs when a high volume of product is discovered missing and determined to be stolen. This type of loss requires that the carrier and manufacturer work in concert with law enforcement to recover the shipment. This is achieved by gathering all pertinent data on the shipment such as product description, serial numbers, and general product information. This information must be passed along to law enforcement immediately if the perpetrators are to be apprehended with the stolen cargo. Some inherent problems exist relative to reporting these thefts to law enforcement so it is imperative to engage all the right people. Although several law enforcement agencies have instituted cargo theft investigative teams, the location of the loss may occur in an area where these resources do not exist. Carriers and manufacturers should work with local, state, and federal law enforcement officials on an action plan in the event of a major loss.

Some of the prominent factors concerning these thefts are that they may occur on nights or weekends from "soft targets." Soft targets are locations where the thief has the ability to enter the location where cargo is stored with minimal chance of detection. It is the responsibility of both the manufacturer and the carrier to provide adequate security measures to minimize significant loss potential. Although the business case for capital improvements to facility security is recognized, thieves will clearly gravitate to the most accessible targets. As the theft is discovered, the state highway patrol and the local police should be contacted. The major loss will have required the perpetrators to have either taken the entire trailer or off-loaded the cargo into another truck. The state highway patrol can quickly determine likely profiles for both vehicle type and likely destinations. Local police can enter the information complete with serial numbers into the National Crime Information Center (NCIC) database, which will provide nationwide intelligence to law enforcement on the theft.

Even though most high-tech cargo thefts represent a significant monetary loss, thefts which do not occur in bulk may be approached from a different perspective. Minor thefts are the result of pilferage, smash-and-grab tactics, or are otherwise not a volume theft. These losses are best investigated by backtracking the entire trail of the shipment. All paperwork that coincides with the shipment should be carefully reviewed. Begin with the customer and the documentation he or she possesses. Serial numbers and BOL exceptions should be immediately reconciled to minimize counting errors or misunderstandings. The customer will have records as to who was responsible for the receiving and unloading of the trailer and those individuals should be interviewed. The carrier is able to retrace the path of the shipment through their network. Dock personnel and drivers can provide history, which should be augmented by paperwork that consists of seal records, trailer manifests, BOLs, and any electronic records that indicate the path of freight movement, including which dock personnel physically handled the shipment. Additionally, where carrier facilities employ the use of guards, records may reflect transit arrival and departure times. Finally, the manufacturer can provide detailed records that show production dates, carrier selection/routing, and who/when was assigned to load the shipment onto the carrier's trailer.

International Organized Crime

There can be no question that the activities of international organized crime have had an effect on legitimate global logistics. Businesses everywhere on the planet can and do feel the negative influences. U.S. businesses, in particular, have been victimized when transporting raw materials and finished goods within the supply chain. Through the billions of dollars that came from their theft operations, the international organized crime cartels have invested their "profits" in legitimate businesses. This laundering of ill-gotten revenue provides the cover necessary to reduce the risks associated with corrupting officials, public enforcement, supply chain employees, and others associated with the transportation industry. The political processes of many countries have been infiltrated or undermined in various forms to further the objectives of organized crime.

Recognizing the impact on the economy and sovereign security through smuggling, terrorism, and supply chain corruption, the United States galvanized all of the federal agencies concerned through the formation of the International Crime Control Strategy (ICCS). The ICCS is designed to bring together all federal enforcement to maximize their effectiveness and make good use of their resources without waste or redundancy.

The goals and objectives of the ICCS are to:

- *Extend the first line of defense beyond U.S. borders.* Basically, this would come from intensifying enforcement and prevention even offshore.
- *Protect U.S. borders by attacking smuggling and related crimes.* Enhance land border inspection and detection through additional resources.
- *Deny safe haven to international criminals.* Accomplish this through new international agreements for prompt identification, arrest, and extradition for international criminals. Create an increase in international enforcement cooperation and sharing of information.
- *Counter international financial crime and money laundering activities.* Target off shore centers and seize the assets of international criminals.
- *Respond to emerging international crime threats* by disrupting new activities of international organized crime groups.
- *Establish international standards to combat crime* through cooperation and intelligence sharing to strengthen the rule of law.
- *Optimize the full range of U.S. efforts* and mobilize and incorporate the private sector into U.S. efforts.

Concerning this last goal, a cooperative effort partnering between government and industry is of significant importance if effective counter-measures and prevention techniques are to be complemented. Making such a joint commitment is essential if we are to improve cargo security within all international corridors. We must make every effort to understand how we interact with the supply chain. The more knowledge we attain, the more effective and successful our investigations will be when we sustain a loss.

Cargo Theft Investigative Techniques

Investigative techniques that can be employed to investigate cargo thefts are limited only by one's imagination. Repeated use of new techniques will enable investigators to become more proficient. Tried and true techniques will always remain in an investigator's bag of tricks.

Some investigative techniques include, but are not limited to the following:

1. Informants

   Although informants were previously discussed as a source of information, the cultivating of informants is an important investigative technique. Informants' personalities vary, as do their motivations. An investigator must be constantly aware of informants' personality traits and motivations without losing sight of the object of the investigation. This can best be accomplished by appropriate direction and control of an informant by the investigator. Informants who come from a cargo related background are diverse.

2. Undercover Investigation

   For many investigators, this is the most popular investigative technique. However, selection of an investigator for an undercover assignment is critical. Cargo situations demand specific needs from an investigator such as knowledge of cargo handling, suitable appearance and physical make up, and possession of a specific mindset.

3. Surveillance

   With the increased mobility of the criminal element, surveillance techniques have expanded from the traditional vehicle and foot to include air, vessel, and hardware. Foot, vehicle, air, and vessel surveillance involve separate skills. Proficiency is acquired only through extensive participation in each type of surveillance.

4. Use of Technical Equipment

   Technical aids which, when properly used, can greatly assist the investigator in his investigation include:

   - Electronic sensors and devices
   - Photography
   - Fluorescent powders
   - Night vision and infrared scopes
   - Closed circuit television with videotape time sequences.

5. Telephone Number Analysis

   Internal and company generated telephone data may be reviewed to determine patterns of activity and possible internal conspirators. A computer system is available for producing reports from toll data in which the information is sorted chronologically and numerically by telephone. This system can be extremely helpful in solidifying cargo theft conspiracy cases that have internal company involvement.

6. Identify "Fence" Operations

   Cargo thefts would be greatly reduced if the thief did not have a readily available outlet to buy the merchandise. Fence operations are directly and indirectly connected with organized crime. Fencing operations vary. In many instances, they operate legitimate businesses or warehouses, but their fencing operations are carried out clandestinely. In recent years, flea markets and swap shops have become a distribution point for fences.

   To identify fences, investigators work undercover and make larger and larger purchases of stolen merchandise. This technique may enable an investigation to proceed in an upward direction in an effort to identify other individuals involved in a theft organization or redistribution process.

7. Storefront Operations

   By identifying fences and making purchases, a certain amount of expertise in this area is gained. The knowledge of a fencing operation and the argot used between a thief and a fence will be of great help when a storefront project is undertaken by public or private enforcements. Storefront operations are time-consuming, expensive, and extremely hazardous, but the payoffs are great when the need for such operations has been identified.

8. Fraudulent Cargo Documentation

   Detecting practices involving fraudulent papers and other forms of misrepresentations that facilitate cargo thefts is a technique that should be developed. Knowledge of documents required for the importation and movement of cargo to its ultimate destination and the authenticity of those documents can disclose fraudulent practices. As soon as fraudulent documentation is used to steal cargo, the authorities should be alerted. The thieves may become aware that their method of operation has been discovered; however, prevention of future thefts is most important.

Cargo Theft Reporting

All reported cargo thefts should be documented. Efforts should be made to determine the extent of investigation required. Some cases of reported theft involve losses so small that investigation, or at least extensive inquiry, may not be warranted. However, investigation of pilferage of even low-value merchandise from security enclosures within bonded premises may be warranted in most cases, since such thefts indicate breakdown in strict controls set up to protect those enclosures where high-value, low-bulk cargo presents an ideal target for thieves. Records are the backbone of an investigation and any effective prevention program.

Every reported theft should be documented. A report prepared in connection with a formal investigation should completely identify the shipment of merchandise and that portion stolen. An indication of actual or estimated retail value should be included in the report. Every effort should be made to include in the report information concerning the merchandise.

1. Name of importing vessel, vehicle, or aircraft.
2. Date and time of arrival of conveyance.
3. Foreign port of lading, if appropriate.
4. Bill of lading number or airway bill number.
5. Identify foreign shipper and/or manufacturer.

6. Identify customs broker (if applicable).
7. Name and identity of ultimate consignee.
8. Carton and/or serial numbers of stolen merchandise (and other identifying marks or numbers).
9. Estimated retail value of stolen merchandise.
10. Indicate whether stolen merchandise is imported or exported.
11. Name and identification of person discovering theft.
12. Location of apparent theft.
13. Date and time theft discovered.
14. Date theft was reported to authorities and to whom.

Identification data for individuals and vehicles who are suspect in the investigation also should be included in the report.

Cargo Documentation Investigations

The movement of cargo is accompanied by documentation from the time it leaves the exporter's premises until it is received by the consignee. To conduct a meaningful investigation, it is necessary to understand the paper flow and the part played by each segment of the industry. Knowledge of general procedures and local differences in these procedures will be extremely beneficial in determining where a theft may have occurred, whether fraudulent documentation was used, and if collusion took place between various elements within the industry.

MARINE CARGO

1. The *steamship company* (or agent) receives via mail a copy of the vessel's cargo manifest. Normally, the consignee is notified two days prior to the ship's arrival. The carrier provides freight release to the terminal operator.
2. The *Customs broker* or consignee obtains freight release, Department of Agriculture clearance, and so forth before contacting the motor carrier. The broker forwards to the motor carrier an original of the domestic bill of lading and the original or the pickup order, which authorizes pick up of the cargo. It is the broker's responsibility to

check the bill of lading for completeness and guarantee with the terminal operator the loading charges and demurrage.

3. The *motor carrier* secures interchange agreement with a steamship company on containers and ascertains expiration of free time and availability of cargo for pick up before dispatching the driver for the pier. The motor carrier furnishes the driver with the original and a copy of the pick up order before his departure for the pier. An appointment is made with the terminal operator at least 24 hours before pick up. Within port movements of bonded shipments (warehouse or general order) require a motor carrier to be a customhouse license holder and require the truckmen to possess customs cartmen identification cards. When bonded shipments are authorized to be transported by common or private carriers, the identification cards are *not* required.
4. The *terminal operator* issues a pass to the driver at the gatehouse. The pick up order is checked for completeness and legibility by the terminal operator dispatcher. The motor carrier's credit rating for loading charges is verified and arrangements made for payment of demurrage, if any. Unless cargo has been authorized for delivery to the steamship company by Customs, the terminal operator directs the driver to the pier Customs office.
5. The *Customs inspector* will verify the driver's papers against prelogged Customs entry documents. When all Customs requirements are completed, the inspector will stamp the delivery order "Delivery Authorized."
6. The *terminal operator* will notify the driver to load and assign a checker and loading area. The cargo is then loaded onto the vehicle by pier personnel and the checker notes exceptions and shortages.

   It is at this point (with inbond entries only) that the joint determination between the checker, acting as the agent of the carrier, and driver, acting as the agent of the importer, present a joint determination to the Customs inspector. If there are any discrepancies, the inspector makes appropriate notations on all copies of the inbond documents, indicating that the joint determination was submitted. The inspector's notation does not affirm or deny the count, it only acknowledges the joint determination and the amount that is permitted. The original delivery order is retained by the terminal operator.

7. It is the *driver's* responsibility to assist in and/or supervise the loading of his vehicle. He signs the delivery order for the quantity received and exceptions and shortages are noted, and retains a copy of the delivery order. Upon departure from the pier, he surrenders the gate pass at the gatehouse. Prior to leaving the pier, the driver notifies the motor carrier dispatcher that he has picked up the cargo and is proceeding with the delivery.

AIR CARGO

1. Upon arrival, an aircraft is given clearance by Public Health and commences to unload its cargo. The *airline* checks the shipment against airway bills and notifies the consignee within 24 hours subsequent to arrival.

2. The *Customs broker* or the consignee prepares the entry. The broker forwards to the inland carrier an original of the airway bill and an original of the pick up order, which authorizes the cargo to be picked up.

3. The *Customs officer* will review all documents and upon examination will authorize delivery of the shipment to the carrier. The Customs inspector may designate a portion of the shipment to be held for appraisal.

4. The *air cargo terminal operator* will examine the pick up order presented by the motor carrier and check it for completeness. If the cargo has not been authorized for delivery by Customs, the motor carrier must obtain this authorization prior to loading.

5. Upon loading of an inbond shipment, the *driver* and *airline representative* will make a joint determination and submit the results on the delivery order to Customs. All imported cargo arriving in the United States, other than noncommercial merchandise carried on the person by a traveler, is temporarily in the custody of the United States Customs Service until the importer or his authorized agent complies with whatever Customs requirements apply. Duty free, noncommercial importations may be cleared to the importer with little or no formality. Whenever formality is involved in the Customs transaction, authorizing delivery of the merchandise to the importer is referred to as an *entry*. Basically, there are

three classes of entry: (1) transportation inbond, (2) consumption, and (3) warehouse.

Inbond Entries

1. Immediate Transportation (IT): This type of entry is used when merchandise arrives at one port, and is released on condition that it be transported inbond to another port. Upon arrival, Customs once again takes temporary custody until an entry of another type is filed by the importer.
2. Transportation and Exportation (T&E): This entry is used when merchandise may be merely passing through the United States to another country. The merchandise is permitted on condition that it be transported inbond out of the United States. Other entries of this type are temporary importation bond and permanent exhibition entry.

Consumption Entries

1. Informal Consumption Entries: These entries are allowed for commercial importations under specified value limits and for noncommercial importations regardless of value.
2. Formal Consumption Entries: At the time a formal consumption entry is filed, the importer pays whatever duties are due on the merchandise. After the shipment is examined and authorized for delivery by Customs, the importer can dispose of the merchandise as he wishes.
3. Immediate Delivery Procedure (ID)—Pending Formal or Informal Entries: This procedure is used to expedite the clearance of cargo. It allows up to ten days for the payment of estimated duty and processing of the commercial entry. In addition, it permits delivery of the cargo prior to payment of estimated duties and allows subsequent filing of the entry.

Warehouse Entries

The use of this type of entry allows the importer to hold his merchandise in a bonded Customs warehouse ready for sale and distribution, but does not have to pay Customs duty until he withdraws the merchandise from the warehouse.

Cargo: Air and Marine Documentation

The movement of goods by air or sea requires proper formal and informal documentation from the time it exits the shipper's facility until it is received by the consignee. To conduct an investigation you must understand the flow of documentation in each segment of movement. Understanding general procedures and any local variances can be helpful in determining where a theft occurred, whether fraudulent documentation was used, and if collusion took place.

Diversion: Grey Goods Investigation and Prevention

Any manufacturer would be wise to screen its product sales and distribution. The acts of diversion that can plague a company are frequently nothing more than white collar crime. Product diversion erodes markets, devalues brand names, destroys a quality reputation or image, negatively impacts on sales, and can create potential liabilities. Certainly, these concerns individually would suggest that customer, distributor, and client screening are a must.

Traditional reliance upon registering and/or obtaining a trademark is not sufficient. While the intent of a trademark is to protect the brand name, it is not adequate to deter diversion and grey market operators. Trade names are protected on a country-by-country basis. While each country protects trademarks in accordance with its own business practices, they are not necessarily registered.

Investigations and prevention efforts should follow the following categories:

- Unfair competition
- Infringement
- Misrepresentation and fraud
- Breach of contract
- Conspiracy to defraud
- Deceptive and unfair business practices
- Dilution of trademark integrity

The U.S. laws that apply may include those listed below, either individually or in combination:

- 18 US Code 1964
- 28 US Code 1331 and 1332
- 18 US Code 1341—Mail Fraud
- 18 US Code 1343—Wire Fraud
- 18 US Code 2314—Transportation
- 15 US Code 1114—Federal Trademark Infringement
- 15 US Code 1125—False Designation of Origin and False Description or Representation to Retailers

Delivery conditions for goods must include that the product is being sold on condition and understanding that it shall be offered for sale in specifically agreed upon locations. The buyer warrants that the product will be delivered to the agreed upon location and remain offered for sale or use at that location. The buyer should furnish a signed bill of lading and copy of the export declaration if originating in the United States. Other appropriate documents used in international logistical transactions may also be required.

Protocols for Grey Goods Control & Diversion

Listed below are ten specific recommendations or protocols which may be implemented by companies for use in transshipment control:

1. All sales and operating employees shall not engage in or support any activities associated with or in support of distribution and sales of its products which results in diversion or transshipments to unauthorized dealers, retailers, or secondary marketers. This shall also apply to suppliers, vendors, or contract employees.
2. Pricing, both domestic and international, should be in the currency of the origin country and uniform in order to reduce or eliminate exchange rate arbitrage.
3. Applications for distributors, retailers, and transporters shall have a due diligence review and business relationship background conducted prior to any agreements for shipments or backgrounds.
4. A task group should be formed by each manufacturer for the purpose of control of product integrity. Membership should consist of representatives of all concerned departments including financial, legal, sales, security, and operations.

5. There should be no approval for shipments or sales disproportionate to realistic abilities of the consignees to sell at retail or distribute.
6. There should be periodic reviews and audits conducted over all domestic and international distributors and retailers. Also included in this process should be all transportation, warehouses, and duty free activities.
7. An agreed upon termination policy should be in place and binding for all customers, clients, and distributors that are proven to be engaged in diversion or transshipment activities.
8. Absolutely no "drop" shipments should be allowed.
9. All large volume and "international" sales should have special markings and packaging.
10. Customers, clients, distributors, retailers, and any other means of product sale shall be given a specific identification, associated only with the individual recipient of the identification.

Investigating Fuel Theft

Both public and private investigators are at times required to investigate the theft of fuel(s), particularly gasoline. This can best be accomplished to the satisfaction of either civil or criminal prosecutions or suits by adding a foreign substance to the fuel/gasoline (one that will not interfere with the operation of a motor or engine). It must be invisible to the naked eye, inexpensive, and easily detected even if the fuel substance is diluted by the addition of untreated fuel. The technique described in this section is attributed to the Enforcement Law and Criminal Investigation for U.S. Treasury Agents.

The necessary material(s) include a 5 percent solution of sodium carbonate. This is available in chemical supply houses, drug stores, and academic and professional laboratories. Never use sodium hydroxide or sodium bicarbonate. A one-half gallon clear container with a sealing cap is also required. An amount of phenolphthalein (not phednolphthalin) is also required, the amount is determined by the amount of fuel/gasoline to be treated for identification purposes. For example, one ounce will treat 10,000 gallons of fuel/gasoline. This substance is relatively inexpensive and readily available.

The following steps must be implemented if the fuel/gasoline is to be identified and traced. First, test the untreated identified fuel/

gasoline by placing 3 pints/1 ½ quarts of untreated gasoline in the half-gallon bottle. Add ½ ounce of sodium carbonate solution. Secure the bottle and vigorously shake the bottle for at least two minutes. Invert the bottle and examine the water layer on the bottom. The water must remain clear and colorless. Empty the bottle and thoroughly wash it.

Second, identify or tag the fuel/gasoline by placing one ounce of phenolphthalein for each 10,000 gallons of fuel in a clear quart bottle. Fill the bottle about three fourths full of alcohol. Seal and shake until the phenolphthalein is dissolved. The contents of the bottle are sufficient to tag or identify 10,000 gallons of fuel/gasoline. It is best to add this solution to the substance to be identified as it is being pumped into the storage tank and also better if it is added in small doses. Obviously, if the storage tank is smaller, say 5,000 gallons, then add one half of the contents of the solution. Caution should be taken to make sure that fuel/gas in the pipeline has been cleaned out so that only the tagged or identified fuel/gasoline is available.

To test for the unauthorized distribution of or possession of any tagged fuel/gasoline, the following steps should be followed:

- Place approximately 2 pints or 1 quart of the suspect fuel/gasoline in a half gallon bottle.
- Add one-half ounce of sodium carbonate solution. Seal the sample and shake vigorously for at least two minutes.
- Invert the bottle and examine the water layer at the bottom. It should be a bluish pink.

Investigating International Cargo Theft

The investigation of losses of cargo in international commerce begins with the understanding that the sources of information and intelligence are determined by the mode of transportation used or the combination thereof. The sources may include the following:

- Shippers.
- Manufacturers.
- Consignees.
- Intermodel transporters.

- Warehouses.
- Container freight stations or break bulk facilities.
- Freight handlers, including forklift and crane operators, lorrie movers, and internal cargo movers.
- Insurance companies.
- Importers/exporters/Customs brokers.
- U.S. Customs Service.
- Federal Bureau of Investigation.
- Federal Maritime Commission.
- U.S. Department of Labor.
- Internal Revenue Service.
- U.S. Department of Transportation.
- State/local law enforcement agencies. This includes air and seaport police, burglary, intelligence/organized crime units, and others as appropriate. Several departments have established specific squads and Task Forces to address the problem (i.e., Los Angeles, Miami, Philadelphia, New York).
- Foreign governments agencies and appropriate authorities from origin countries.
- Appropriate councils, organizations, associations: for example, American Trucking Association, Air Transport Association, Waterfront Commission of New York Harbor, Association of Marine Underwriters, international associations, sea and airport police, and the National Cargo Security Council.

The United States Customs Service Office of Investigations

Cargo Security Initiatives

The effective operation of the U.S. economy depends on the efficient movement of people, goods, information, and financial resources. In the United States, it is estimated that more than three million people are employed in the freight transportation industry. In 1998, more than 20 million containers were imported into or exported out of the

United States. Government estimates indicate that the value of imports into the United States will nearly triple by the year 2002.

Cargo theft-related crime is a serious problem that affects all modes of transportation. Theft-related crime disrupts the reliable and efficient flow of goods and is very expensive. Theft-related losses, which include the direct cost of stolen cargo, higher insurance premiums, and additional administrative expenses, reduce transportation industry profits and increase prices for consumers. The extent of cargo theft in the United States goes unnoticed by most people outside of the transportation industry or the law enforcement community. Annual cargo theft losses are estimated to be in the billions of dollars. In fact, some estimates put cargo theft-related losses in the United States at $10 billion a year and $30 billion worldwide. Gathering accurate estimates of cargo theft losses is difficult due to limited reporting by the transportation industry and the lack of a national law enforcement system for uniform reporting and tracking. Ironically, these deficiencies not only hinder law enforcement efforts to identify, arrest, and prosecute individuals and criminal organizations engaged in cargo theft-related crime, but they also ensure that law enforcement resources devoted to cargo theft remain limited.

Criminal organizations targeting the transportation industry are a serious and ever increasing problem. Airports, seaports, carriers, warehouses, and other freight facilities with inadequate security practices and procedures provide a haven for individuals and criminal organizations engaged in cargo theft, drug and contraband smuggling, money laundering, the exportation of stolen vehicles and merchandise, and a host of other illegal activities.

U.S. Customs Service investigations have demonstrated that the majority of cargo theft-related crime is committed by, or with the collusion of, persons whose employment gives them access to the cargo or merchandise that is stolen or pilfered. Employees use their position and access to engage in the theft of merchandise. Their familiarity with scheduling, routing practices, transportation personnel, and Customs Service procedures has provided them with the necessary knowledge to successfully carry out these crimes. Perhaps more alarming is that many of these same employees may also be part of international criminal conspiracies engaged in major drug smuggling and money laundering activities. Internal conspiracies are criminal conspiracies that involve either airline, shipping, security, trucking, freight forwarding, warehouse, or brokerage personnel in attempts to steal merchandise or import controlled substances or other contraband through the movement of legitimate international cargo shipments. The scope of internal conspiracies is vast. Over the last 30 years, internal conspiracies have been documented and shown to exist in airport and seaport operational areas throughout the United States and around the world.

As the nation's principal border agency, the mission of the U.S. Customs Service is to ensure that all goods and persons entering and exiting the United States do so in accordance with all United States laws and regulations. The Customs Service is the only federal agency with statutory and regulatory authority to enforce smuggling laws in the United States and is the only agency with statutory authority to conduct warrantless searches and seizures at the border or the functional equivalent (airports or seaports).

The Customs Service regulates the entire importing and exporting process, including customs warehouses and container stations. Customs brokers, bonded carriers, cartage carriers and lighters/barges, centralized examination stations, and foreign trade zones. All imported merchandise is in the Customs Service's custody until it authorizes or permits the carrier to make delivery to the consignee or the next carrier. Customs bonded warehouses, foreign trade zones, and container stations are required by customs regulations to immediately report any theft or suspected theft of merchandise to the U.S. Customs Service port director. Any loss or detention of bonded merchandise, or any accident involving a vehicle or lighter carrying bonded merchandise must be immediately reported by the cartman, lighterman, bonded carrier, foreign trade zone operator, bonded warehouse proprietor, container station operator, or centralized examination operator to the U. S. Customs Service port directory. Additionally, international carriers, bonded carriers, container stations, and centralized examination stations are responsible and accountable for any merchandise which is manifested or for which the bonded carrier or other custodian has formally receipted. Therefore, thefts or suspected thefts from these carriers should also be reported to the U.S. Customs Service port director.

The Customs Service enforces laws for a multitude of illegal activities pertaining to imported and exported merchandise and contraband, including the theft of merchandise from Customs Service custody. The United States Customs Service is committed to the prevention and detection of cargo theft, as well as aggressively investigating those individuals and criminal organizations involved in cargo theft. Special agents (criminal investigators) from Customs' Office of Investigations, are located in over 130 domestic field offices throughout the United States and in 27 foreign countries. Special agents conduct all criminal and civil investigations involving violations of customs and related laws and have recovered millions of dollars worth of stolen cargo.

Augmenting the investigative and enforcement efforts of the Customs Service are partnerships and initiatives between Customs and manufacturers, shippers, carriers, and port authorities. Through these joint public and private efforts, the Customs Service is dedicated to assisting foreign and domestic ports and the trade community in enhancing their security programs. These industry partnership programs afford the Customs Service the opportunity to share "best" security practices with the international trade community. Some of these programs include the Carrier Initiative Program (CIP), the Business Anti-Smuggling Coalition (BASC), and the Americas Counter Smuggling Initiative (ACSI).

The Carrier Initiative Program started in 1984 as a result of a substantial increase in the smuggling of cocaine and marijuana in south Florida. It is a joint effort among air, sea, and land carriers and the U.S. Customs Service. The goal of CIP is to prevent and deter the smuggling of drugs on commercial conveyances. This is done by providing carriers with an incentive to initiate security improvements in their cargo facilities, on their conveyances, and to enhance their drug smuggling awareness. In February 1999, approximately 3,800 air and sea carriers and 975 land border carriers participated in the CIP.

International carriers should be aware of the enormous monetary penalties they may incur if the Customs Service discovers unmanifested drugs on board a conveyance. Recent amendments to federal laws have significantly increased monetary penalties against carriers who fail to manifest drugs upon importation into the United States. By raising the drug manifesting penalties, the U.S. Congress intended that carriers play a greater role in stemming the flow of drugs in commercial shipments. Under current law, a carrier must satisfy its legal burden of proof that it did not know and could not have known by the exercise of the highest degree of care and diligence that unmanifested drugs were on board its conveyance. The Customs Service will look for evidence that a carrier took meaningful steps to deter or prevent drugs from being smuggled on board its conveyance. Through the administrative petitioning process, the Customs Service may mitigate a penalty in response to a carrier's diligent and good faith efforts to comply with the law. The Customs Service will consider whether a carrier is a signatory to a carrier initiative agreement and, more significantly, will look to the degree of implementation of the agreement when contemplating mitigation. Finally, the Customs Service permits penalty offsets (a dollar-for-dollar match) for the purchase and installation of "extraordinary" security equipment (e.g., remote cameras, x-rays, etc.). A carrier must, of course, submit sufficient evidence to support its claims. It is the Customs Service's view that penalty dollars would be more beneficial if the carrier invested the money in such extraordinary security equipment.

In March of 1996, the Business Anti-Smuggling Coalition was formed between the trade community and the U.S. Customs Service. BASC is designed to complement and enhance the CIP program. This concept is a business-led, Customs-supported alliance created to combat narcotics smuggling via commercial trade and to promote a more security conscious environment. There are more than 200 companies currently participating in BASC, with official BASC chapters now established in six cities in South America.

The Americas Counter Smuggling Initiative was developed as a priority undertaking by U.S. Customs in 1997. The goal was to build on the successes of CIP and BASC by strengthening and expanding the antinarcotics security programs with industries and governments throughout Central and South America. Targeted countries include Venezuela, Colombia, Peru, Mexico, Panama, Costa Rica, and Ecuador. Over the past three fiscal years, CIP and BASC participants provided information which resulted in more than 450 domestic and foreign seizures, totaling more than 110,000 pounds of illegal drugs. During the 1998 fiscal year, participants in CIP and BASC affected 82 foreign intercepts totaling 12,665 pounds of illegal drugs, and provided information to the U.S. Customs Service resulting in 54 domestic drug seizures totaling 21,217 pounds. Although the main focus of these industry partnership programs is to safeguard the movement of legitimate freight, the added security practices instituted by the freight industry act as a deterrent to many cargo-related criminal activities.

As an example of a Customs Service port security initiative, Customs, in concert with the Jacksonville (Florida) Port Authority, the U.S. Coast Guard, and three major cargo carriers, is currently involved in creating a model port security project. This project is intended to combat the smuggling of narcotics into the United States via commercial cargo, and to create minimum security standards that could serve as a model for other ports throughout the United States. It will be demonstrated that a sound security plan incorporated into the daily operations of ports will be advantageous to all interested parties, in that security investments are good business practices that will pay commercial dividends in the long term.

Why have we discussed anti-drug smuggling initiatives in a publication that is devoted to cargo security and loss control? The answer is simple. The same types of security measures instituted to prevent and deter drug smuggling activity are also highly effective against cargo theft-related crime.

Companies can obtain assistance from the Customs Service with improving their security measures to prevent cargo theft-related crime by contacting the U.S. Customs Service, Office of Field Operations, Industry Partnership Programs (see Appendix B).

Part Eight

SECURITY SURVEYS AND AUDITS

SECURITY SURVEYS AND AUDITS

A security survey is a critical on-site examination and analysis of an industrial plant, business, home, or public or private institution. It is used to ascertain the present security status, to identify deficiencies or excesses, to determine the protection needed, and to make recommendations to improve the overall security of the site.

It is interesting to note that a definition of crime prevention as outlined by the British Home Office Crime Prevention Program—"the anticipation, recognition and appraisal of a crime risk and the initiation of action to remove or reduce it"—could, in fact, be an excellent definition of a security survey. The only difference, of course, is that a survey generally does not become the "action" as such but rather a basis for recommendations for action.

This definition can be divided into five component parts and analyzed so that its implications can be applied to the development of a working foundation for the security surveyor.

The Anticipation

How does the anticipation of a crime risk become important to the security of *crime*-prevention surveyor? Obviously, one of the primary objectives to a survey is the anticipation or prevention aspects of a given situation—the pre- or before concept. Thus, an individual who keeps anticipation in the proper perspective would be maintaining a proper balance in the total spectrum of security surveying. In other words, the anticipatory stage could be considered a prognosis of further action.

Recognition

What means will provide an individual who is conducting a survey the relationships between anticipation and appraisal? Primarily, the ability to recognize and interpret what he perceives to be a crime risk becomes one of the important skills a security surveyor acquires and develops.

Appraisal

The responsibility to develop, suggest, and communicate recommendations is certainly a hallmark of any security survey.

Crime Risk

This, as defined in this text, is the opportunity from crime. The total elimination of opportunity is most difficult, if not most improbable. Thus, the cost of protection is measured in

1. Protection of depth.
2. Delay time.

Obviously, the implementation of the recommendation should not exceed the total (original or replacement) cost of the item(s) to be protected. An exception to this rule would be human life.

The Initiation of Action to Remove or Reduce a Crime Risk

This section indicates the phase of a survey in which the recipient of the recommendations will make a decision to act, based on the suggestions (recommendations) set forth by the surveyor. In some cases the identification of security risk is made early in a survey, and it is advisable to act upon the recommendation prior to the completion of the survey.

The responsibility to initiate action based on recommendations is the sole duty of the recipient of the survey. This is to suggest that the individual who receives the final evaluation and survey will be the individual who has commensurate responsibility and authority to act.

There are basically three types of surveys: *A building inspection* is advising a tenant in a large complex of his vulnerabilities as they pertain to the physical characteristics of the dwelling. *A security survey*, on the other hand, would be conducted on the whole complex versus doing only a portion of the site. *A security analysis* is more of an in-depth study including risk management and analyzing risk factors, environmental and physiological security measures, crime patterns, and evidence of fraud and internal theft.

THE BEST TIME TO CONDUCT THE SURVEY

Most crime-prevention officers and security directors agree that a survey is most effective

1. after a crisis within the cooperation;
2. after a breaking and entering or major larceny; and
3. upon request.

There are times when a merchant hears that he can get something for nothing and thereby calls the crime-prevention officer in the town to conduct such a survey, when in reality he has no intention of spending a dime for improvement. A close friend of mine conducted a detailed security survey on a factory warehouse and office building. The recipient of the survey followed only one of his recommendations, which was to leave a light on over the safe in the back room of his warehouse. The owner had completely disregarded other recommendations such as hardware improvements on doors, windows, and skylights. Unfortunately, thieves returned and almost put him out of business.

DEVELOPING YOUR SECURITY POINTS

Like most professionals, you need tools to do an effective job. The following are suggested to assist you when conducting your surveys: tape measure, floor plans, magnifying glass, flashlight, camera with flash, small tape recorder, screwdriver, penknife, pencil, and paper.

Your survey should be conducted systematically so that the recipient can follow your recommendations in some kind of order. Start with the perimeter of the building. Once inside the building, start at the basement and work your way to the attic. Do not be afraid to be critical of the area that you are in. This is what the recipient wants. After you have done several surveys, you will develop a style of putting them together and they become easy.

NINE POINTS OF SECURITY CONCERNS

1. General purpose of the building, i.e., residence, classroom, office. Consider the hours of use, people who use the building, general hours of use, people who have access, key control, and maintenance schedule. Who is responsible for maintenance? Is the building used for public events? If so, what type and how often? Is the building normally opened to the public? Identify the significant factors and make recommendations.

2. Hazards involving the building or its occupants. List and prioritize, i.e., theft of office equipment, wallet theft, theft from stockrooms. Identify potential hazards that might exist in the future.

3. Police or guard security applications. What can they do to improve the response to the building and occupants from a patrol, investigation, or crime-prevention standpoint? Would the application of guards be operationally and/or cost effective?

4. Physical recommendation. Inspect doors, windows, lighting, access points. Recommend physical changes that would make the building more secure, such as pinning hinges on doors and fences.

5. Locks, equipment to be bolted down, potential application of card control and key control. Make specific recommendations.

6. Alarms. Would an alarm system be cost effective? Would the use of the building preclude the use of an alarm? Are the potential benefits of an alarm such that the building use should be changed to facilitate the use of an alarm? Consider all types of alarms, building-wide or in specific offices. Consider closed-circuit television application and applications for portable or temporary alarm devices.

7. Storage. Are there specific storage problems in the building, i.e., expensive items that should be given special attention, petty cash, stamps, calculators, microscopes? Make specific recommendations.

8. Are there adequate "No Trespassing" signs posted? Are other signs needed?

9. Custodians. Can they be used in a manner that would be better from a security standpoint?

COMMUNICATION

The first step in conducting a security survey is to communicate with the recipient of the survey. This meeting will serve as your focal point for reviewing the recipient's risk factors and environmental and physiological security measures, as well as give you an opportunity to discuss any past losses, crime patterns, fraud, internal theft, and past larcenies that have occurred.

The meeting with upper as well as middle management is important because these people can provide valuable input to the designing of your report. Remember that these individuals are the decision makers, and they are the ones who will be supporting and reinforcing your recommendations and long-range changes that are to be implemented. Getting support from the top is like getting a commitment. Once you have obtained that support, the rest is easy.

At times you may find that in-house security personnel may be having a hard time communicating to upper management that their security is poor and defenses are weak. You therefore become the middle managers' intermediary. Make sure that your actions and recommendations support your communications.

There should be at least two meetings with management. The first meeting is to define your objectives, goals, and scope of the inspection. The final meeting, when the report is finished, is to describe your findings in detail and go over your recommendations.

CRIME ANALYSIS

It is not necessary for you to be a statistician, but the more you know about and understand the local crime problems, the better equipped you will be to analyze the potential crime for risk loss in surveying a business or a home. Crime-analysis collection is simply the gathering of raw data concerning reported crimes and known offenders. Generally, such information comes from crime reports, arrest reports, and police contact cards. This is not to say that these are the only sources available to collect crime data. The analysis process as applied to criminal activity is a specific step-by-step sequence of five interconnected functions:

1. Crime data collection.
2. Crime data collation.

3. Data analysis.
4. Dissemination of analysis reports.
5. Feedback and evaluation of crime data.

These functions give you the *risk factor* upon which you set the degree of security that is required for the facility.

Crime Statistics

Detailed information that pinpoints the types and locations of a jurisdiction's or organization's crime problems is the key to the design of a prevention program. The following discussion thus focuses on such statistics. Burglary is becoming a crime of epidemic proportion in this country, and the outline of data elements relates directly to that crime. For those instances where a jurisdiction or organization is also interested in robbery, etc., similar types of data about those crime types will have to be pulled together for program-design purposes.

When to Collect Information and Why

Crime statistics must be gathered as a first step in the design of a prevention program. In addition, they will have to be maintained throughout the program. In fact, they should be maintained on a daily basis and summarized monthly. This information will serve as one of the baselines for designing, managing, and evaluating the program. In addition, the data will provide useful details for inclusion in speeches, news stories, and briefing memorandums.

What Information to Gather

The following data elements should be considered essential in the design of a burglary-prevention program. The degree to which a particular organization will be able to gather such data will, of course, depend on the nature of existing records systems. The *essential* elements include:

Location Occurrence

Census tracts and exact locations of crimes being analyzed should be gathered. This information will be valuable in identifying program target areas.

Date of Occurrence

The day, month, and year of incidents, or reporting dates if exact dates are not known, should be analyzed. These dates will indicate whether particular crimes are related to seasonal patterns, weekly community habits, or to close-downs of plant operations, thus enhancing burglar opportunities. This information will be useful in formulating strategies to deal with the crime.

Time of Occurrence

The exact time of occurrence, when known (or morning, afternoon, evening, when a range of time is given) should be analyzed. This information will be helpful to the security unit in terms of making personnel-allocation plans for example.

How Entry Was Made

Record what was done to get inside the facility. This might include whether force was used; if a door or window had been left open and was utilized; whether a lock and/or alarm system was defeated to gain entry; etc. This data will be valuable to the formulation of public information and target-hardening strategies.

Type of Property Taken

Information on the types of items taken, such as money, sound or office equipment, etc., should be developed. Keep such accountings general in nature. This information will be helpful in the design of target-hardening strategies.

Property Identification

Record whether property taken was identified with special markings. This will be valuable in designing a program as well as in evaluating its impact.

Dollar Amount Taken

Record the total reported dollar loss of all items taken. This data can be used to help analyze the impact of your program, for example, the dollar value loss before and after the implementation of a program.

Property Damage Resulting from Entry

Note the dollar losses in property damage resulting from the crime. This information can be utilized to evaluate the impact of the program, for example the dollar loss through property damage before and after implementation of a program.

How Incident Was Reported

Gather data on who detected and reported the crime, such as an employee, police, alarm, victim, etc. This will be useful in designing a program; crime prevention programs are theoretically designed to increase citizen and employee participation and reporting. Baseline statistics have to be developed to see if this actually happens.

In addition to the data elements noted above, there are various other items that may be useful in the design, monitoring, and evaluation of a program. However, these items should be considered optional. They may, nonetheless, be gathered if time is available; the reporting systems used by the organization include the items; information is available through a computer or data-processing system; and so on. As such, each unit will have to consider its own situation to determine whether the following items should be gathered. Such items include

Point of Entry

Record where entry was made; such as door, window, air vent, etc. As appropriate also note the location of the point of entry, such as rear door, or side window.

Occupancy Status

Record whether the premises were occupied when the reported crime occurred.

Instrument Used to Gain Entry

Record data on the kind of tool or instrument used, such as screwdriver, pry bar, saw, etc. should be gathered.

Street and Exterior Lights

For night crimes, note the existence of street and exterior lights.

Points of Entry Lighting

For night crimes, record the existence of lights at the point of entry.

Visibility at Points of Entry

Record whether the point of entry was visible to patrolling units.

Alarm Systems

Record the existence of alarm systems.

Status of the Alarm

Record whether the alarm was activated or defeated.

When Detected

Record whether the crime was detected while in progress, on the day of occurrence, the next day, etc.

THE REPORT

If you have done everything mentioned in the preceding section, then in all probability you have completed your survey. However, keep in mind the following pitfalls.

1. Do not over-exaggerate in your report. Management will see through such exaggeration.
2. Do not inflate your report with unnecessary maps and floor plans, thereby giving the appearance of writing a book rather than a report. Use such aids only when a specific point is to be illustrated.
3. Do not repeat your statements over and over or on page after page.
4. Do not jump around in your report, for example, from perimeter protection to roof recommendations, second-floor coverage, and basement openings. Place your recommendations in order of first

the perimeter, then from the basement up to the roof, thereby allowing the recipient an opportunity to follow in your footsteps systematically floor by floor.

5. Do not be afraid to be critical. This is your objective and this is what the recipient expects.

SURVEY FOLLOW-UP EVALUATION

The final step to your survey is to evaluate the effectiveness of your survey after the recipients have reviewed it and implemented security changes that have been recommended. A question that you might be asking yourself is: "Why didn't they do this or that as compared to what they actually did do?" In all probability the following would be appropriate responses.

1. Did you make your recommendations clearly?
2. Did you identify the specific objectives that your client should be addressing?
3. Did you consider the total physical and overall structure of the site, keeping in mind that each building has its own personality and life style?
4. Did you properly plan and communicate with management your overview before leaving the site?
5. Did you properly communicate in your report your client's vulnerabilities and specific needs.

After you have followed management guidelines in doing your security survey, you should inquire as to how completely your ideas were accepted. This should be done in a very low-key way.

PREPARING THE SECURITY SURVEY REPORT

A. The Facility

   1. Name and address of the facility
   2. Location: City or town, zip code
      a. Proximity to other facilities
      b. Proximity to communities
      c. Telephone number

3. Physical nature of the facility and its surroundings
4. Function, number of departments within the facility
5. Staffing
    a. Nature and size of staff
    b. Working hours
    c. Seasonal variations
    d. Number of people who have access (security officers, cleaners, etc)
    e. Who does maintenance
    f. Maintenance schedule
    g. Clearance level
6. Visitor access
    a. Visitor use characteristics
    b. Number of visitors
    c. Hours, type of weekend operation
    d. Seasonal variations
7. The climate
8. The law-enforcement situation
    a. Type of jurisdiction
    b. Reaction time
    c. Distance from nearest fire station
    d. Law-enforcement powers of security staff
9. Specific security problem of site
10. Identify area of highest dollar value

B. Perimeter Security (External)

1. Fences and natural barriers
    a. Nature of fences and gates
    b. Gates (control, access, locks, etc.)

      c. Natural barriers
      d. Clear zones
   2. The grounds (includes roads and parking areas)
   3. Lighting
      a. Natural
      b. Location and areas of coverage
      c. Maintenance and testing
      d. Power supply and circuit reliability
      e. Tamper resistance
      f. Operation
   4. Access control
   5. Areas of cover for hiding and materials that would aid someone in gaining entry to the area
   6. Intrusion-detection system
      a. Type of system
      b. Signal transmission mode
      c. Inspection, testing, and maintenance
      d. Power supply
      e. Tamper resistance
      f. Records of alarm incidents
   7. Patrols
      a. Nature
      b. Frequency
      c. Seasonal Variations

C. The Building
   1. Perimeter security
      a. Doors
      b. Windows

   c. Other openings
   d. Walls
   e. Roofs
   f. Floors and basement
2. Interior security
   a. Connecting doors and pass-throughs
   b. Walls
   c. Ceilings
   d. Ductwork
   e. Store rooms and closets
   f. Vaults
   g. Storage cabinets
   h. Hiding places
   i. Restrooms
3. Locks and related hardware, hinges, padlocks, and guard plates
4. Lighting
   a. Security
   b. Emergency
   c. Lighting reliability, interior and exterior
      (1) Back-up power
      (2) Secure switches
      (3) Maintenance
5. Intrusion-detection system
   a. Nature and area of coverage
   b. Location of controls
   c. Tamper resistance of wiring and components
   d. Alarm transmission mode

e. Maintenance and testing

f. Back-up power

g. Operating procedures and instructions

6. Fire safety and control

a. Detection systems-ask the same questions asked for Intrusion-detection system above

b. Suppression modes

c. Housekeeping

d. Evacuation

7. Safes

D. Procedures

1. Key control (see separate checklist)

2. Building opening, closing, and search

3. Housekeeping practices

4. Employee control

a. Screening and investigation

b. Identification and pass control

c. Package and material control

5. Visitor control

a. Control of visitors to staff-only areas

b. Passes

c. Records or visits

d. Ids for contractors, tradesmen, utility workers, etc.

6. Visitor (public) surveillance and inspection

7. Protection of administrative records

8. Security of cash valuables

a. On-site

b. In transit

9. Control of access to restricted areas of facilities
10. Property inventory and control
11. Security communications
    a. Modes
    b. Reliability
    c. Back-up power and alternate modes of communication
    d. Employee operation
    e. Efficiency and speed
12. Incident reporting
    a. Timeliness
    b. Accuracy

E. Emergency Action Plan and Contingency Plans
   1. Possibly contingencies
      a. Bomb threats and bombings
      b. Civil disturbances (riots, war, breach of peace)
      c. Natural catastrophes
      d. Vandalism
      e. Robbery and assault
      f. Burglary
      g. Fire
      h. Equipment failure
      i. Accidents with injury
   2. Notification of the right people in the right sequence
   3. Detailed reaction instructions or directions on how to get instructions
   4. Reporting of emergency incidents to the regional office and to the proper local authorities

F. Storage Area Protection

1. Store rooms
    a. Physical construction
    b. Access control
    c. Environmental control
    d. Housekeeping practices
    e. Storage cabinets and shelves
    f. Inventory and material movement
2. Furnished rooms
    a. Access control
        (1) Physical
        (2) Procedural
    b. Alarm
    c. Environmental controls
    d. Housekeeping practices
    e. Inventory

PERIMETER PROTECTION

The Exterior

The purpose of a strong exterior perimeter is to reduce the risk, defeating the criminal before he gets inside. Several years ago, museums were being entered from the skylights and priceless art was stolen. A hole in the perimeter security went undetected by the administration of the museum but not by the thief.

Many devices are available to protect the perimeter by means of alarming your chain-link fence, or enclosing this fence within barbed-wire fencing or by means of optic fibers placed underground, under water, and within glass windows. These are not complex systems, but just basic coverage. However, we also have aesthetics to deal with as well as the implementation of security devices.

The perimeter protection program is designed to:

1. Detect and verify any attempt for unauthorized entry.
2. Delay and slow down the individual, while appropriate assistance is arriving.
3. Deter, or prevent individual access. This requires studying the overall environment. All sites have a personality; day, night, and weekend staff all should be examined.
4. Deny his objective to the thief. Your objective is to prevent losses.

Emphasis on the environment of the site is important in order to construct the most effective security program.

Barriers

The most basic protection for cargo is a barrier that prevents the would-be thief from approaching his cargo target. Physical and mechanical barriers and recommended procedures to be followed outside the perimeter of the terminal facility will be discussed in this chapter.

Types of Barrier

Barriers can be used to create physical and psychological deterrents to accidental entry; to prevent deliberate unauthorized entry; to delay intrusion, making detection and apprehension by guards more likely; to make guards more effective; and to direct the flow of pedestrian and vehicular traffic.

There are two kinds of physical barrier—natural and structural. Natural barriers include rivers, marshes, or terrain difficult to negotiate on foot or by vehicle. Structural barriers include fences, walls, buildings, grills, bars, and gates. A barrier should be under physical or electronic surveillance to be fully effective.

The kind of barrier to be used depends upon the size of the controlled area, the flow of traffic during the busiest and least busy periods, and the most prevalent local hazards.

The perimeter of a large controlled area may be a combination of natural and structural barriers. A limited area, however, must generally be bounded by a structural barrier.

Fencing

Fences should be chain link, no. 9 gauge or heavier wire; no less than eight feet high, with mesh openings no larger than 2 inches per side and with a twisted, barbed selvage at the top. It should be stretched taut and securely fastened to metal posts set in concrete. The bottom should be within 2 inches of hard ground or paving. On soft ground, it should extend below the surface to compensate for shifting soil or sand. Culverts, troughs, or other openings larger than 96 square inches in area should be protected by fencing or iron grills to prevent unauthorized entry, yet allow proper drainage.

A top guard should be attached to perimeter fences and interior enclosures for greater security. A top guard is an overhang of barbed wire along the top of a barrier facing outward and upward at an angle of 45 degrees. The supporting arms, at least 2 feet long, are attached to the tops of the fence posts. At least 3 strands of standard barbed wire are tightly stretched between the supporting arms. Some fences have a double overhand, facing both outward and inward, which makes it more difficult to enter or leave the facility by scaling the fence.

The top guard can be firmly fixed or mounted on springs. The spring-type guard further increases the difficulty of scaling the fence. If a building less than three stories high forms part of the perimeter, a top guard should be used along the coping to deny access to the roof.

The fence line should be as straight as possible to provide ease of observation by the guard force. If practicable, fences should be located no closer than fifty feet to buildings or cargo in a controlled area. Twenty feet of clearance should be allowed between the perimeter barrier and exterior features such as buildings or parking areas that would offer concealment to a thief.

Fencing for limited areas should conform to the same specifications as for controlled areas, although it is recommended that the height be increased and that a top guard be used.

Exclusion areas should be located in secure buildings; they consist of separate cribs, cages, or vaults. If fencing is used, it should be at least ten feet high, extend to the ceiling or be topped by a wire mesh roof, and be under observation by a guard.

If a wall serves as the barrier, or a part of it, it should be constructed to provide protection equal to that specified for each of the areas above.

If it is less than the height specified, it should be topped with chain-link fence and barbed wire to match the minimum requirements. If a fence connects with a building, it should extend within two inches of the wall of the building.

A body of water, whether it be river, lake, or ocean, does not in itself constitute an adequate barrier. Additional measures, such as a fence, frequent security patrol, and floodlighting usually are necessary.

To be effective, barriers have to be well maintained. Breaks or damage to the structure should be repaired as soon as they are discovered. Frequent inspections of the barriers must be made by the guard force to locate defects. In addition, the security officer should periodically tour the barriers, giving particular attention to cuts or openings in the barrier, which may be camouflaged.

If the perimeter barrier encloses a large area, an interior all-weather road should be provided for guard-force vehicles. The road should be in the clear zone and as close to the barrier as possible. Its use should be limited to guard and emergency vehicles.

Entrances

The number of gates and entrances to controlled areas should be limited to the minimum required for safe and efficient operation of the facility. A top guard equal to that on the adjoining fence should be attached to each gate. The bottom of the gate should be within two inches of hard ground or paving. Adequate lighting should be provided for fast, efficient inspection.

When gates or doors are not manned by guards so that all those entering will be challenged, they should be securely locked, illuminated during hours of darkness, and periodically inspected by a roving guard.

Semi-active entrances, such as railroad siding gates, or gates and doors used only during peak traffic-flow periods should be locked except when actually in use. Keys to these entrances should be in the custody of the security officer or the chief of the guard force and should be strictly controlled. Periodic inspection should be made of these entrances.

Inactive entrances, which are used only occasionally, should also be kept locked. They are subject to the same key control and inspection as semi-active entrances.

Emergency exits should have alarmed break-out hardware installed on the inside.

Sidewalk elevators and other unusual entrances that provide access within controlled-area barriers should be locked and patrolled.

Control signs stating the condition of entry to a facility or controlled area should be erected at all entrances. They should inform the entrant that he is subject to search of his person (subject to the extent the law allows), vehicle, or packages, and they should clearly state any prohibitions against packages, matches, smoking, or entry for other than business. The signs should be legible under normal conditions at least fifty feet from the point of entry.

To maintain the integrity of the barriers to controlled areas, guard-control stations should be established at all entrances in service.

PERIMETER SECURITY CHECKLIST

The checklist below may be used for many different types of facility. It permits each facility manager to select those elements pertaining to his establishment and location in making his own security survey.

Barriers

1. Is the perimeter of the facility or activity defined by a fence or other type of physical barrier?
2. If a fence or gate is used, does it meet the minimum specifications?
   a. Is the top guard strung with barbed wire and angled outward and upward at a 45 degree angle?
   b. Is it at least 10 feet in total height?
   c. Is it located so that it is not adjacent to mounds, piers, docks, or any other aid to surmounting it?
3. If building walls, floors, and roofs form a part of the perimeter barrier, do they provide security equivalent at least to that provided by a chain-link fence? Are all openings properly secured?
4. If a masonry wall or building forms a part of the perimeter barrier, does it meet minimum specifications of perimeter fencing?

5. If a river, lake, or other body of water forms any part of the perimeter barrier, are security measures equal to the deterrence of the 10-foot fence provided?
6. Are openings such as culverts, tunnels, manholes for sewers and utility access, and sidewalk elevators, which permit access to the facility, properly secured?
7. List number, location, and physical characteristics of perimeter entrances.
8. Are all portals in perimeter barriers guarded, secured, or under constant surveillance?
9. Are all perimeter entrances equipped with secure locking devices, and are they always locked when not in active use?
10. Are gates and/or other perimeter entrances that are not in active use frequently inspected by guards or management personnel?
11. Is the security officer responsible for security of keys to perimeter entrances? If not, which individual is responsible?
12. Are keys to perimeter entrances issued to other than facility personnel, such as clearing, trash-removal, or vending-machine service personnel?
13. Are all normally used pedestrian and vehicle gates effectively and adequately lighted so as to ensure
    a. proper identification of individuals and examination of credentials,
    b. that interiors of vehicles are clearly lighted, and
    c. that glare from luminaries is not in guards' eyes?
14. Are appropriate signs setting forth the provisions for entry conspicuously posted at all principal entrances?
15. Are clear zones maintained for the largest vehicles on both sides of the perimeter barrier? If clear-zone requirements cannot be met, what additional security measures have been implemented?
16. Are automobiles permitted to park against or too close to the perimeter barrier?

17. What is the frequency of checks made by maintenance crews of the condition of the perimeter barrier?
18. Do guards patrol perimeter areas?
19. Are reports of inadequate perimeter security immediately acted upon and the necessary repairs effected?
20. Are perimeters protected by intrusion-alarm devices?
21. Does any new construction require installation of additional perimeter barriers or additional perimeter lighting?

LIGHTING AND SECURITY

What would happen if we shut off all the lights at night? Stop and think about it!

The results of such a foolish act would create an unsafe environment. Senior citizens would never go out, and communities would have an immediate outbreak of thefts and vandalism. Commercial areas would be burglarized at an uncontrollable rate. Therefore, lighting and security go hand in hand.

The above example may seem to be farfetched, but in fact, installation of improved lighting in a number of cities has resulted in the:

1. decrease in vandalism,
2. decrease in street crimes,
3. decrease in suspicious persons,
4. decrease in commercial burglaries, and
5. a general reduction in crime.

If the above principles are true, then you have no alternative but to take a hard look at your facility's lighting program.

AT NIGHT

At night a protective lighting system enables your guard force to maintain a level of security approaching that observed during the day. Adequate lighting is relatively inexpensive. If it cannot be provided, management has to consider other more costly alternatives, such as additional guards, sentry-dog patrols, or expanded alarm systems.

The amount and intensity of light needed will vary from point to point within the facility. Designing a system for a large facility is a specialized task. Material is available from the manufacturers of lighting equipment that will assist management, but consultation with an expert in the field will save time and expense. It will undoubtedly produce a more satisfactory lighting system.

Protective lighting will permit guards to observe activities around or inside a facility. It is achieved by providing even light on areas bordering the facility, directing glaring light into the eyes of a potential intruder, and maintaining a low level of light on guard patrol.

Planning

When planning a protection lighting system, the creation of high contrast between an intruder and the background is a primary consideration. The ability of a guard to distinguish a darkly clothed man against a dark background improves significantly as the level of illumination is increased. Predominantly dark, dirty surfaces require more light to facilitate observations than do those of clean concrete or light-colored paint. This is also true inside buildings, where ceilings and walls redirect and diffuse light.

Generally, lighting should be directed downward and away from the structure or area to be protected, and away from the guards assigned to patrol the facility. It should create as few shadows as possible.

It must be stressed that recommendations that are made regarding lighting should not, whenever possible, work contrary to operational needs. Further, common sense is the key requirement when making any recommendations or suggestions as to type, number, and placement of lighting fixtures. Historically, operational management of a terminal is constantly looking for additional space; where containers are concerned it often stacks them three high directly in front of a "field" of light from a light fixture, which renders that fixture ineffective. Other objects often create problems similar to these such as debris, pallets, field cargo, and equipment, which is stacked and/or stored carelessly and which obstruct fields of light.

In order to avoid any wrong interpretations of recommended locations of light fixtures, detailed diagrams should be drawn of the perimeter locations and structures showing the exact recommended locations.

When recommending locations, much depends on the type of lighting fixture and distance between placements. Importance must be placed upon the type and condition of interior lighting as well as of exterior lighting. Having well-lighted loading platforms, clerical offices, and storage areas is just as essential as having adequate perimeter lighting. The need to be able to easily read documents necessary to the movement of cargo is imperative. Clerical errors often provide a means by which cargo theft can be perpetrated.

Consideration for special-area lighting should be given special attention, such as for

1. guard houses,
2. gates,
3. parking areas,
4. truck lines,
5. scale houses,
6. string pieces and aprons,
7. secure areas such as cargo cribs, and
8. specific trouble areas where known and repeated entries have been made.

Under almost all circumstances, it should be recommended that the use of automatic timers or photoelectric cells be employed on outdoor lighting. This is necessary to remove the responsibility of turning on and off protective outdoor lighting systems from one or more individuals.

A planned maintenance program to retain the protective lighting system as near to 100 percent effective as possible is always a necessary recommendation. Even some otherwise efficient terminal operators become lax regarding the replacement of bulbs, fixtures, and other lighting materials when necessary. In addition, include providing protection to the lighting fixtures. Bulbs are a favorite target for vandals; they can be easily protected from damage by wire mesh or some other substantial material that would not restrict the light beam.

Do not expect security or safety lighting installations or upgrading to solve all your security problems. What you can expect is to greatly improve the overall effectiveness of your security force and operational management's ability to *see* what it is that is under their control and supervision. Just remember that security and operational lighting is only one of the many methods that a professional security manager can use to help provide proper asset protection.

Good principles of protective lighting are achieved by adequate light upon bordering areas, glaring light in the eyes of the intruder, and no light on the guard or guard. When you wish to provide protective seeing, high brightness contrast between the intruder and the background should be a number-one consideration. This can be accomplished by adequate light and the use of proper light colors when painting surfaces. You can use two approaches to provide this contrast, or a combination of the two. The first method is to light the boundaries and approaches; the second method is to light the area and structures within the general boundaries of the terminal or property.

LIGHTING CHECKLIST

1. Is the perimeter of the installation protected by adequate lighting?
2. Are the cones of illumination from lamps directed downward and away from the facility proper, and away from guard personnel?
3. Are lights mounted to provide strips of light both inside and outside the fence?
4. Are lights checked for proper operation periodically, and are inoperative lamps replaced immediately?
5. Do light beams overlap to provide coverage in case a bulb burns out?
6. Is additional lighting provided at vulnerable or sensitive areas?
7. Are gate guard boxes provided with proper illumination?
8. Are light-color finishes or stripes used on lower parts of buildings and structures to aid guard observation?
9. Does the facility have a dependable auxiliary source of power?
10. Is there alternate power for the lighting system independent of the plant lighting or power system?
11. Is the power supply for lights adequately protected? How?
12. Is the standby or emergency equipment tested periodically?

13. Is emergency equipment designed to go into operation automatically when needed.
14. Is wiring tested and inspected periodically to ensure proper operation?
15. Are multiple circuits used? If so, are proper switching arrangements provided?
16. Is wiring for protective lighting securely mounted?
    a. Is it in tamper-resistant conduits?
    b. Is it mounted underground?
    c. If above ground, is it high enough to reduce the possibility of tampering?
17. Are switches and controls properly located, controlled, and protected?
    a. Are they weatherproof and tamper resistant?
    b. Are they readily accessible to security personnel?
    c. Are they located so that they are inaccessible from outside the perimeter barrier?
    d. Is there a centrally located switch to control protective lighting? Is it vulnerable?
18. Is the lighting system designed and locations recorded so that repairs can be made rapidly in an emergency?
19. Is adequate lighting for guard use provided on indoor routes?
20. Are materials and equipment in shipping and storage areas properly arranged to permit adequate lighting?
21. If bodies of water form a part of the perimeter, does the lighting for them conform to other perimeter lighting standards?

LOCKS AND KEY CONTROL

Locks are an essential and integral part of barriers and the security they provide. To be effective, however, keys and combinations to locks must be strictly controlled. If these are compromised, the security of the entire facility is compromised.

Regardless of their quality or cost, locks can be considered only as delay devices. They are not positive bars to entry. Many ingenious locks have been developed, but equally ingenious means have been devised to open them surreptitiously. Some locks require considerable time and expert manipulation to open, but even they will eventually succumb to force and the proper tools. The protection afforded by a well-constructed lock can be measured in terms of the time the locking mechanism will resist picking, manipulation, or drilling.

To determine a facility's lock requirements demands specialized knowledge. Few security officers, and fewer transportation managers, have the necessary expertise. It is advisable, when selecting the equipment to be used, to consult a professional locksmith.

Locks commonly used in the transportation system include these types.

1. Key locks, some of which can be opened by an expert in a few minutes. The ease with which a key may be lost and compromised or with which an impression may be made should be considered in determining the security value of a key lock.
2. Conventional combination locks, which may be opened by a skillful manipulator able to determine the settings of the tumblers of a common three-position dial type through his sense of touch or hearing. Although some combination locks may require several hours to open, a skillful person can open the conventional combination lock in a few minutes.
3. Manipulation resistant combination locks, which are designed so that the opening lever does not come in contact with the tumblers until the combination has been set. This type provides a higher degree of protection for important material.
4. Other combination locks with four or more tumblers that afford still greater protection for very important items.
5. Relocking devices, which furnish an added degree of safety against forcible entry to a safe or vault door. These devices increase the difficulty of opening a combination lock by punching, drilling, or blocking. They are recommended for heavy sales and vaults.
6. Interchangeable-core locks, which have cores that can be removed, replaced by others using different keys. The cores can be replaced quickly, instantly changing the matching of locks and keys if their

security is compromised. Another advantage is that all the locks in a facility can be keyed into an overall master-key system. Interchangeable-core locks are economical, involving lower maintenance and new-lock expenses. The system simplifies record-keeping, is flexible, and can be engineered to the needs of the facility.

7. Kingpin locks, which are placed on the kingpin of a trailer or container chassis to make it impossible to connect a tractor. They provide a medium to high degree of security, depending on the type of lock. Although expensive, one lock can protect a number of trailers simultaneously if the locked trailer is parked in a blocking position. Because the lock can provide positive security, it is recommended for use in an area where only periodic surveillance, rather than constant observation is possible.

Key Control

For effective control of locks, keys, and combinations, accurate records must be maintained and periodic physical inventories made. Combinations and keys should be issued only to those whose official duties require them.

Combinations to safe locks and padlocks securing containers should be changed at least once during each twelve-month period; immediately following the loss or possible compromise of a combination or key; after the discharge, suspension, or reassignment of anyone who knows the combination; or upon receipt of a new container with a built-in combination lock.

Key and combination control for the facility should be exercised by the security officer. Records containing combinations and keys should be securely stored, with only limited and controlled access allowed. Lists of persons authorized to draw keys to controlled areas should be kept in the key-storage container. Key containers should be inventoried at the end of each shift and all keys accounted for. Above all, keys should not be issued for permanent retention or removal from the facility. Keys should be logged out at the beginning of each shift and logged in at the end. The following list summarizes guidelines for key control.

1. Purchase a large key cabinet to store and control the many keys that are in your possession.

2. Two sets of keys should be furnished or obtained with the new key cabinet. Both keys in each set should be identified with tags.

    a. One tag should read: *file-key, must not be loaned out.*

    b. The second tag should read: *Duplicate.*

    The key cabinet should be equipped with "loan tags" that will identify the person to whom a key is loaned.

    These tags, when needed, are hung on numbered pegs corresponding to the keys that are in use.

3. Establish accurate records and files, listing the key codes, the date a key was issued, and who received it.
4. Have each employee sign a receipt when he or she receives a key.
5. All alarm keys should be marked and coded.
6. A check should be made of what keys are in the possession of watchmen and staff.
7. Do not issue keys to any employee unless absolutely necessary.
8. Only one person should order and issue keys for the complex.
9. Change the key cylinder when an authorized key holder is discharged for cause. Furthermore, terminated or retired employees should produce keys previously issued at the time of termination.
10. Periodic inspections should be made to assure that possession of keys conforms to the record of issuance. These periodic inspections should be utilized to remind key-holders that you should be immediately notified of any key loss.
11. The original issue of keys and subsequent fabrication and reissuance of keys should assure that their identity is coded on the keys so that the lock for which they were manufactured cannot be identified in plain language.

Each facility will have requirements for key and lock control systems peculiar to itself. A survey should be conducted to determine the actual need for additional protection afforded by locking devices. When this determination has been made, an annex to the security plan can be drafted, showing

1. The location of key depositories.
2. The keys (by building, area, or cabinet number) to be held in each depository.
3. The method of marking or tagging keys for easy identification.
4. The method of control for issue and receipt of keys, including register maintenance and identification of personnel authorized to receive keys.
5. The action required if keys are lost, stolen, or damaged.
6. The frequency and method of lock rotation.
7. The assignment of responsibility and accountability by job or position title.
8. The availability of emergency keys to the guard supervisor.
9. A list of persons to whom this plan is made available.

KEY CONTROL AND LOCK SECURITY CHECKLIST

1. Has a key-control officer been appointed?
2. Are locks and keys to all buildings and entrances supervised and controlled by the key-control officer?
3. Does the key-control officer have overall authority and responsibility for issuance and replacement of locks and keys?
4. What is the basis for the issuance of keys, especially of master keys? Are keys issued only to authorized personnel? Who determines who is authorized? Is the authorization in writing?
5. Are keys issued to other than installation personnel? If so, on what basis? Is it out of necessity or merely for convenience?
6. Are keys not in use secured in a locked, fireproof cabinet? Are these keys tagged and accounted for?
7. Is the key cabinet for duplicate keys regarded as an area of high security?
8. Is the key or combination to this cabinet maintained under appropriate security or secrecy? If the combination is recorded, is it secured?

9. Are the key-locker and record files in order and current?
10. Are issued keys cross-referenced?
11. Are current records maintained indicating
    a. buildings and/or entrances for which keys are issued?
    b. number and identification of keys issued?
    c. location and number of master keys?
    d. location and number of duplicate keys?
    e. issue and turn-in of keys?
    f. location o flocks and keys held in reverse?
12. Is an audit ever made, asking holders to actually produce keys, to ensure that these keys have not been loaned or lost?
13. Who is responsible for ascertaining the possession of keys?
14. Is a current key-control directive in effect?
15. Are inventories and inspections conducted by the key-control officer to ensure compliance with directives? How often?
16. Are keys turned in during vacation periods?
17. Are keys turned in when employees resign, are transferred, or are fired?
18. Is the removal of keys from the premises prohibited when these keys are not needed elsewhere?
19. Are locks and combinations changed immediately upon loss of theft of keys or transfer or resignation of employees?
20. Are locks changed or rotated within the installation at least annually regardless of transfers or known violations or key security?
21. Are current records kept of combinations to safes and the dates when these combinations are changed? Are these records adequately protected?
22. Has a system been set up to provide sub-masters to supervisors and officials on a "need" basis, with facilities divided into different zones or areas?

23. If master keys are used, are they devoid of marking identifying them as master keys?
24. Are master keys controlled more closely than change keys?
25. Must all requests for reproduction or duplication of keys be approved by the key-control officer?
26. Are key-holders ever allowed to duplicate keys? If so, under what circumstances?
27. When the manufacturer's serial number on combination locks and padlocks might be visible to unauthorized persons, has this number been recorded and then obliterated?
28. Are locks on inactive gates and storage facilities under seal? Are seals checked regularly by supervisory or key-control personnel?
29. Are measures in effect to prevent the unauthorized removal of locks on open cabinets, gates, or buildings?
30. Are losses or thefts of keys and padlocks promptly reported by personnel and promptly investigated by key-control personnel?
31. If the building was recently constructed, did the contractor retain keys during the period when construction was being completed? Were locks changed since that time? Did the contractor relinquish all keys in his possession during construction after the building was completed?
32. If removable-core locks are in use, are unused cores and core-change keys given maximum security against theft, loss or inspection?
33. Are combination-lock, key, and key-control records safeguarded separately (i.e., in a separate safe or file) from keys, locks, cores, and other such hardware?

Are all locks of a type that offers adequate protection for the purpose for which they are used?

INTRUSION ALARMS

Intrusion-detection systems should be used as a supplement to, but not as a substitute for, your facility's guard force. The alarm is intended to alert guards of an intrusion or attempted intrusion. The results depend upon the response time of the guards and the quality and effectiveness of the alarm system. Before an intrusion-detection system is installed, you should conduct a threat or risk analysis. Knowing the risk is the primary prerequisite for designing an effective security system. However, in drafting the design of an alarm system, this system should be treated for what is actually is: an alert for well-trained security-guard forces.

Alarm systems come in all shapes and sizes, each having a specific purpose and installed a specific way. In alarming your museum, library, art gallery, corporate office, or home containing a considerable amount of art, *every exterior door, window, skylight, or other opening must be secured and alarmed.* Additional interior coverage can be either spot protection or saturation of structure. These systems consist of

1. Perimeter protection devices: magnetic contacts, switches, foil, broken-glass detectors, pressure mats, and accessories.
2. Space-protection devices: photoelectric, ultrasonic, microwave, passive infrared, and audio detectors.
3. Object protection devices: capacitance proximity detectors, pressure switches, vibration transducers, safe contacts.
4. Audible annunciation: horns, sirens, bells, rotating beacon lights. (A silent on-site alarm may be preferred).
5. Alarm systems sending signals to one or more of the following: central station, police connect, tape of digital dialer, proprietary or local.
6. Control panels, with U.L. listing, custom made or computerized, AC or DC with or without power backup, with or without high-security shunt keys.

These systems are affected by temperature changes, humidity, metal objects, animals, water movement, thin walls, heaters, the sun, ventilation ducts, and curtains.

INTRUSION ALARM SECURITY CHECKLIST

1. Is an alarm system used in the facility?
    a. Does the system indicate an alert only within the facility?
    b. Does it signal in a central station outside the facility?

c. Is it connected to facility guard headquarters?

d. Is it connected to directly to an enforcement headquarters outside the facility proper? Is that headquarters a private protection service? A police station? A fire station?

2. Is there any inherent weakness in the system itself?
3. Is the system supported by properly trained, alert guards?
4. Is the alarm system for operating areas turned off during working hours?
5. Is the system tested prior to activating it for non-operational periods?
6. Is the alarm system inspected regularly?
7. Is the system tamper resistant? Weatherproof?
8. Is an alternate alarm system provided for use in the event of power failure?
9. Is the emergency power source designed to cut in and operate automatically?
10. Is the alarm system properly maintained by trained personnel?
11. Are periodic tests conducted frequently to determine the adequacy of response to alarm signals?
12. Are records kept of all alarm signals received to include time, date, location, action taken, and cause for alarm?
13. Are quality and effectiveness of the correct alarm system examined?
14. Are all motion-detection units and all areas that have magnetic contacts or switches periodically walk-tested?
15. Is the system experiencing false alarms, and has the cause been determined?
16. How frequently is the system completely inspected by who, and is a report filed?
17. Can parts be obtained within eight hours, thereby allowing for minimum down time?
18. Is more than one vendor capable of servicing the system?

19. Does the system have panic buttons, and are these buttons tested frequently?
20. Are all alarm keys accounted for?

ADDITIONAL CHECKLISTS

INSPECTION REPORT FORMS

The effective security director will never rest on his laurels after completing his security plan and training curriculum. He will realistically take note of the fact that problems within the industry change from time to time. A constant updating of his security program can best be accomplished by occasionally conducting an inspection. The following form, developed during the Cargo Theft Program at the National Crime Prevention Institute, Louisville, Kentucky, by consultant Harlan Flinner, is an excellent tool for such purposes.

The thoughtful security director needs to constantly monitor all areas of his overall security plan. This means setting up a checklist of possible problem areas that he foresees as troublesome. As a result of this forethought, many security directors have a tendency of developing a form of "tunnel vision," and often miss problems developing in so-called "safe" areas. A checklist of the type we suggest is of great importance to pick up on areas that are too easily overlooked. Setting a regular schedule to cover this checklist is all that is necessary; everything else follows logically. Potential problem areas are being monitored on a regular basis, and small isolated areas are routinely covered. If "vigilance" is the watchword of all security, this check list is an excellent systematic method of maintaining that vigilance.

TERMINAL SECURITY INSPECTION REPORT

____________________ Terminal ____________________ Date

____________________ Inspected by

PURPOSE: A periodic security check should be made of each terminal to determine whether or not standard security and operating procedures are being followed: to check the handling of freight; the vulnerability of the terminal to theft by outsiders; the condition of the fencing and lighting; to provide a vehicle for evaluating the overall security of the terminal; and to make possible specific recommendations.

A. Terminal Building exterior

1. Condition of the building ____________
2. Outside doors ____________
3. Windows clean and in good repair? ____________
4. Lighting ____________
5. Company identification sign ____________
6. Condition of parking area ____________
7. Employee and guard cars parked in proper parking area? ____________

B. Terminal Building (Interior)

1. Reception area for visitors? ____________
2. Condition of offices ____________
3. Condition of store rooms and supply rooms ____________
4. Arrangements for janitorial/cleaning services ____________
5. Condition of restrooms ____________
6. Terminal Office ____________

C.

1. What are bank deposit arrangements and/or are they satisfactory? ____________

   ____

   ____
2. How are paychecks distributed? ____________
3. Checks and drafts secured at all times? ____________
4. Petty cash locked up? ____________
5. Is a safe available? In use? ____________
6. Are keys assigned? ____________

   ____
7. Are keys redeemed if key personnel leave? ____________

   ____

8. Are locks changed as warranted? ______________________
9. D/R file ______________________
10. Term 64 file ______________________
11. Term 51 file ______________________
12. Pud manifest file ______________________
13. Summary manifest file ______________________
14. Photo file ______________________
15. Fingerprinting equipment ______________________
16. Fingerprinting policy compliance ______________________
17. Polaroid camera ______________________
18. Photographic policy compliance ______________________

D. City Drive Check-In

1. Non-Delivery Report (Term 31) used properly? ______________________

____

2. Over/Damage Report (Term 31) used properly? ______________________

____

3. Do bills of lading reflect driver's signature and number of pieces picked up? ______________________
4. Tally sheets attached to bills in the morning? ______________________

____

5. Tally sheet attached to bills when they re turned in? ______________________

____

6. Exceptions being listed in accordance with company policy?

____

E. OS&D Records

1. Shortage files being properly maintained? ______________________
2. "ALL SHORT" bills filed in pre-twx file? ______________________

3. "ALL SHORT" Bills reported on (Term 80) Control Report? ____

____

4. Security Reports made to Regional Security Chief? ________

____

5. Over freight processed per policy? ________________

6. Any bills over 30 days old on the Control List? Why? ________

____

7. Refused shipments properly safeguarded? ____________

____

F. Dock Area

1. What is general appearance of dock? ______________

____

2. Is the dock striped? ______________________

3. Are doors and bays clearly identified? ______________

4. Adequate storage for OS&D freight? ______________

____

5. Hot room (or cage) available and in use? ____________

6. Are any dock lights burning during daylight hours? ________

____

7. Are dock door locks in good condition? ______________

____

8. Fire extinguishers available? Inspected? ____________

9. Re-cooping equipment available? ______________

10. Are there any damaged shipments on the dock? ________

____

11. Are OWB shipments written up promptly and placed in OS&D bay? ________________________

12. Seal procedures
    a. Seal procedure being followed? ______________________

    ____

    b. Are seals adequately controlled? ______________________

    ____

    c. Discrepancies reported to supervisors ______________
13. General condition of trailers and loads during this inspection

    ____

    ____

    ____

    ____

    ____

    ____

14. Trash receptacles available? In use? Need emptying? _______

    ____

15. Reward posters displayed on bulleting board? ___________

    ____

16. Are locks assigned to city units? Policy being followed? _____

    ____

G. Freight Bills Procedures
   1. Bill boxes available on dock? ______________________
   2. Freight bills handled in accordance with established procedures? ______________________________________

      ____

   3. Does break-out man account for all bills given him? _______

      ____

4. Are "security shipments" given proper treatment? (firearms, clothing, etc.)

5. Are shortages, overages and damaged freight brought to foreman's attention?

H. Guard Service

1. Hours of service?
2. Written reports submitted by guard?
3. Are there written instructions for the guard?
4. Is the service adequate? Specific problem areas?
5. Is the guard performing other than security-related duties?
6. Is supervision by guard company adequate?
7. Are complaints processed to a satisfactory and speedy solution?
8. Are emergency phone numbers available?

I. Yard Area

1. Yard fenced?
2. Fence condition?
3. Grass kept cut on either side of fence?
4. Are gates locked at night?
5. Are high value loads protected?

6. Signs posted on or near gates prohibiting cars from entry? ___

   ___
7. Is employee access controlled? ___
8. Lighting
   a. Adequate? ___
   b. Controlled by timer? ___
   c. Any burned out? ___
   d. Would additional lights help? ___
9. General condition of yard ___

   ___

   ___

   ___

   ___

   ___

   ___

J. Maintenance Area

1. Gas pumps locked when the terminal is shut down? ___

   ___
2. Fire extinguishers available? Inspected? ___
3. Maintenance work area orderly? ___
4. Appear well organized? ___
5. Parts and tools properly secured? ___
6. Inventory method used? ___

This report was discussed with ___ by ___

at ___ on ___ 19___

Security Inspection Report Master Log

Year:

Terminal	Region	Date of Inspection

PHYSICAL SECURITY CHECKLIST

1. Do you have a written security plan that is updated at least once a year?
2. Are the functions of security clearly defined?
3. Do you have a preplanned procedure for internal and external notification in case of theft or emergency?
4. Do you have fixed management responsibility for theft or pilferage?
5. Are your records stored and safeguarded properly?
6. Does your shipping and receiving facility have adequate lighting?
7. Is your entire shipping and receiving area fenced or enclosed?
8. Are your entrance and exit gates secured properly or are they secured only by the use of chains and padlocks?
9. Do you have a sufficient number of competent checkers for your needs?
10. Do you have a sufficient number of supervisors?
11. Do you check your facility during the evening and early morning hours to determine the caliber of supervision during those times?
12. Is there a pass system for logging vehicles in and out?

13. Do you have a security crib for placement of high-value cargo?
14. Where called for, are interior or exterior alarms used?
15. Is the monitoring of your alarms provided on a continual basis?
16. Do you have padlocks issued to equipment operators for use when vehicles are left unattended?
17. when assigning keys to individuals, are signatures obtained?
18. Do you issue master keys only where operationally essential?
19. Are periodic checks conducted on key control and locking devices?
20. Are spare keys and locking devices kept under tight controls?
21. Do you check to see if equipment parked or left unattended is left with keys in the ignition?
22. Is there a pre-designated high-value holding area for cargo held in shipping and receiving areas?
23. Do you periodically examine the refuse removal and cleaning service equipment and personnel when they leave your facility?
24. Have you screened the vendor companies with which you are doing business?
25. Are you conducting a pre-employment investigation on all prospective applicants?
26. Are you taking fingerprints and/or photographs of all new employees?
27. Are casual employees included in your pre-employment screening process?
28. Is there a probationary period for new employees?
29. Are your personnel records safeguarded?
30. Have you displayed reward posters with proper wording and considerations?
31. Are you engaged in a security education program?
32. Is your total management team security conscious?
33. Are you providing any type of in-service training for dealing with security problems?

34. Is your paper documentation controlled?
35. Are you tallying cargo on and off equipment properly?
36. Are you using seals and recording the numbers?
37. Are seal numbers being verified at destination?
38. Do you keep your seals under tight security controls?
39. Are seals being affixed by responsible management personnel?
40. Is there a report made to security for trailer or container seal discrepancies if found?
41. Are files maintained which indicate the number of times an employee is involved in a theft or shortage?
42. If dealing with interline activities, are controls maintained on their movements and equipment checked?
43. Do you restrict your employees, including management, from parking near cargo operations?
44. Are controlled parking areas provided for employee's cars?
45. Do you prevent unauthorized employee access to areas where cargo or paper flow dealing with cargo is being handled?
46. Are unauthorized personnel and unidentified visitors allowed in your shipping and receiving areas?
47. Are employees prohibited from visiting their vehicles during working hours?

COMMUNICATIONS CHECKLIST

1. Is the security communications system adequate?
2. What means of communications are used?
    a. Telephone
        (1) Is it a commercial switchboard system? Independent switchboard?
        (2) Is it restricted for guard use only?
        (3) Are switchboards adequately guarded?

(4) Are there enough call boxes and are they conveniently located?

(5) Are open wires, terminal boxes, and cables frequently inspected for damage, wear, sabotage, and wire-tapping?

(6) Are personnel cautioned about discussing cargo movements over the telephone?

b. Radio

(1) Is proper radio procedure practiced?

(2) Is an effective routine code being used?

(3) Is proper authentication required?

(4) Is the equipment maintained properly?

c. Messenger. Is the messenger always available?

d. Teletype. Is an operator available at all times?

e. Public address.

(1) Does it work?

(2) Can it be heard?

f. Visual signals.

(1) Do all guards know the signals?

(2) Can they be seen?

3. Is security communications equipment in use capable of transmitting instructions to all key posts simultaneously?

4. Does the equipment in use allow a guard to communicate with guard headquarters with minimum delay.

5. Is there more than one system of security communications available for exclusive use of security personnel?

6. Does one of these systems have an alternate or independent source of power?

7. Has the communications center been provided with adequate physical security safeguards?

PERSONNEL IDENTIFICATION AND CONTROL CHECKLIST

1. Is an identification card or badge used to identify all personnel within the confines of the controlled areas?
2. Is the identification medium designed to provide the desired degree of security?
3. Does the identification and control system include arrangements for the following:
   a. Protection of the meaning of coded or printed components of badges and passes?
   b. Designation of the various areas requiring special control measures to which the badge holder may be authorized entrance?
   c. Strict control of identification data?
   d. Clear explanation and description of the identification data used?
   e. A clear statement of the authorization and limitations placed upon the bearer?
   f. Details of where, when, and how badges shall be worn?
   g. Procedures to be followed in case of loss or damage to identification media?
   h. Procedure for recovery and invalidation?
4. If a badge exchange system is used for any restricted area, does the system provide for:
   a. Comparison of badge, pass, and personnel?
   b. Physical exchange of restricted area badge for general authorization badge at time of entrance and exit?
   c. Logging a record of each badge exchanged?
   d. Inventory of badges issued by security personnel at the start and completion of tours of duty?
   e. Location of personnel who have not checked out of the area at the close of each tour of duty?
   f. Security of badges not in use?

5. Are messengers who are required to traverse areas of varying degrees of security provided with special identification?
6. Are the prescribed standards for access to exclusion areas supplemented with arrangements and arrangements for the following:
   a. At least one representative of management or security is in the area at all times when work is in progress?
   b. No other persons are permitted to enter the area until one representative of management or security has entered?
   c. A representative of management or security remains until all others have departed?
7. Are personnel, who require infrequent access to a critical area and who have not been issued regular security identification for the area, treated as "visitors" thereto, and issued either (1) a visitor's badge or pass, or (2) a special pass?
8. Are all personnel required to wear the security identification badge while on duty?
9. Do guards at control points compare badges to bearers both upon entry and upon exit?
10. Is supervision of personnel charged with checking identification badges sufficient to insure continuing effectiveness of identification and control system?
11. Are badges recorded and controlled by rigid accountability procedures?
12. Are lost badges with one bearing a different number or one that is otherwise not identical to the one lost?
13. Are procedures relative to lost, damaged, and/or forgotten badges adequate?
14. Are temporary badges used?
15. Are lists of lost badges posted at guard control points?
16. Are badges of such design and appearance as to enable guards and other personnel to recognize quickly and positively the authorizations and limitations applicable to the bearers?
17. How long ago were currently use badges originally issued?

18. Do existing procedures insure the return of identification badges upon termination of employment?
19. Are badges similar to or identical to employee badges issued to outside contractor employees working within the installation?
20. Have local regulations governing identification and control been revised in any material respect since first established?
21. Are all phases of the system under supervision and control of a security officer?
22. Is an effective visitor escort procedure established?
23. Are visitors required to conspicuously display identification on outer garments at all times while on installation?
24. When visitors leave the installation, are they required to turn in their identification badges, and is the departure time in each case recorded on the visitor's register?
25. What procedures are involved when visitor identification badges are not turned in prior to departure of the visitor?
26. Is there a central receptionist?
    a. If "yes," specify functions.
    b. Are functions performed under the supervision of a security officer?
27. Are receptionists (or guards) stationed at different focal points to maintain visitor control?
28. Are there special procedures applicable to visitors requiring access to cargo handling documents?
29. Are special visitors, e.g., vendors, tradesmen, utility servicemen or special equipment servicemen, issued a special distinctive type of visitor badge.
30. What measures are employed, other than the issuance of identification badges, to control the movements of personnel from other transportation companies working within the perimeter of the facility?
31. Does the system used for identification of truck drivers and helpers conform with security regulations?

32. Is the security officer the single responsible official for all aspects of visitor control?

PACKAGE AND MATERIAL CONTROL CHECKLIST

1. Is there standard procedure on control of packages and materials?
2. Are all guards conversant with the package control measures?
3. Are notices on restriction and control procedures prominently displayed at each active entrance and exit?
4. Is there a checkroom where employees and visitors can leave their packages?
    a. Is an adequate receipt system in effect?
    b. Are packages inspected in the owner's presence before a receipt is issued?
    c. Is access to the checkroom restricted to authorized personnel only?
    d. Is a policy established for disposition of items left beyond a specified period?
5. Are spot checks of persons and vehicles conducted and, if so, are frequency and scope thereof indicated?
    a. Regular search
    b. Spot search
    c. Special search
6. Are detection devices used?
    a. X-ray or other similar device
    b. Metal detector
    c. Other; evaluate effectiveness
7. Is a property removal slip, signed by an authorizing official, required when property is being removed from the facility?
8. Are removal slips available in the security office for signature by officials authorizing property removals?
9. Are property removal slips surrendered to guards at exit points?

10. Are special rules established for package and material handling?
    a. Is package and material pass used to exempt bearer from search?
        (1) Is time, date, bearer's name, using agency, and description of the contents, properly recorded thereon?
        (2) Is preparation and issue rigidly controlled?
        (3) Is it serially numbered?
        (4) Does it provide for signature of validating officials?
        (5) Is signature card readily available to guards for comparison?
    b. Is a trustworthy and identified courier used at all times?
11. Are special clothing issued for wear in the facility to prevent the introduction or removal of unauthorized items?
12. Is an effective procedure used for control and search of special vehicles?
    a. Emergency vehicles
    b. VIP vehicles
    c. Special courier vehicles
    d. Vendor's vehicles
    e. Vehicles with loads which are impracticable to search
13. Is there close coordination between security headquarters and the activities that handle cargo movements?
14. Are new employees given appropriate instructions relative to the handling and safeguarding of cargo?

VEHICLE CONTROL CHECKLIST

1. Are vehicles which are allowed regular access to the facility registered with the security officer?
2. Have definite procedures been established for the registration of private cars and are they issued in writing?

3. Do the vehicle registration requirements apply also to motor vehicles owned or operated by employees of any individual firm, corporation or contractor whose business activities require daily or frequent use of vehicles on the facility?
4. Is annual or more frequent registration required?
5. What information is incorporated in registration application forms?
6. Do the prescribed prerequisites for registration include a valid state registration for the vehicle and a valid state operator's license?
7. Is mechanical inspection of vehicles and/or proof of financial responsibility required as a prerequisite of authority to operate a vehicle within the facility?
8. Are decalcomania or metal permit tags affixed to all vehicles authorized to operate within the facility?
9. Do registration permits bear a permanently affixed serial number and numerical designation of year of registration?
10. Do the regulatory controls for registration include:
    a. Prohibition against transfer of registration permit tags to use with a vehicle other than the one for which originally issued?
    b. Replacement of lost permit tags at the registrant's expense?
    c. Return of tags to the security officer when the vehicle is no longer authorized entry into facility?
    d. Destruction of invalidated decalcomania or metal tags?
11. What is the nature and scope of registration records maintained by the security officer?
12. Do the gate guards make periodic checks to insure that vehicles are operated on the premises only by properly licensed persons?
13. Is a specified system used to control the movement of commercial trucks and other goods conveyances into and out of the installation area?

14. Are loading and unloading platforms located outside the operating areas, separated one from the other, and controlled by guard-supervised entrances?
15. Are all trucks and other conveyances required to enter service gates manned by guards?
16. If trucks are permitted direct access to operating areas, are truck drivers and vehicle contents carefully examined?
17. Does the check at entrances cover both incoming and outgoing vehicles?
18. Are truck registers maintained?
19. Are registers maintained on all company vehicles entering and leaving the facility?
20. Are escorts provided when vehicles are permitted access to operating or controlled areas?
21. Does the supervision of loading and unloading operations insure that unauthorized goods or people do not enter or leave the installation via trucks or other conveyances?
22. Are company trip tickets examined?
23. Is a temporary tag issued to visitor's vehicles?
24. Are automobiles allowed to be parked within operating or controlled areas?
25. Are parking lots provided?
26. Are interior parking areas located away from sensitive points?
27. Are interior parking areas fenced so that occupants of automobiles must pass through a pedestrian gate when entering or leaving the working area?
28. Are separate parking areas provided for visitors' vehicles?
29. What is the extent of guard surveillance over interior parking areas?
30. Are there restrictions against employees entering private vehicle parking areas during duty hours?
31. Are automobiles allowed to park so close to buildings or structures that they would be a fire hazard or obstruct fire fighters?

32. Are automobiles permitted to be parked close to controlled area fences?
33. Are parking facilities adequate?

GUARD FORCES CHECKLIST

1. Is a guard force provided? Is it responsive to management authority?
2. Indicate authorized and actual strength, broken down by positions.
3. Have there been changes since the last survey in either the authorized or actual guard force strength?
4. Is present guard force strength commensurate with the degree of security protection required?
5. Is the use of guard forces reviewed periodically to assure effective and economical use?
6. Is supervisory responsibility for guard force operations vested in the security officer?
7. Is a guard headquarters provided?
8. Does the guard headquarters contain control equipment and instruments of all alarm, warning, and guard communications system?
9. Are guards familiar with the communications equipment used?
10. Does the guard headquarters have direct communication with local municipal fire and police headquarters?
11. Do members of the guard force meet the minimum qualifications standards?
12. Are guards armed while on duty? If so, with what type of weapon?
13. Are the weapons kept in arms racks and adequately secured when not in use?
14. Are ammunition supplies properly secured and issued only for authorized purposes?

15. Is each member of the guard force required to complete a course of basic training and take periodic courses of in-service or advanced training?
16. Are the subjects included in the various training courses adequate? Does the training cover:
    a. Hand-to-hand combat?
    b. Care and use of weapons?
    c. Common forms of pilferage, theft, and sabotage activity?
    d. Types of bombs and explosives?
    e. Location of hazardous materials and processes?
    f. Location and use of fire protective equipment, including sprinkler control valves?
    g. Location and operation of all important steam and gas valves and main electrical switches?
    h. Conditions which may cause fire and explosions?
    i. Location and use of first aid equipment?
    j. Duties in the event of fire, explosion, natural disaster, civil disturbance, blackout or air raid?
    k. Use of communication system?
    l. Proper methods of search?
    m. Observation and description?
    n. Patrol work?
    o. Supervision of visitors?
    p. Preparation of written reports?
    q. General and special guard orders?
    r. Authority to see force, conduct searches, and make arrests?
17. Are periodic examinations conducted to insure maintenance of guard training standards?
18. Are activities of the guard force in consonance with established policy?

19. Is supervision of the guard force adequate?
20. Are general and special orders properly posted?
21. Are guard orders reviewed at least semiannually to insure applicability?
22. Are periodic inspections and examinations conducted to determine the degree of understanding and compliance with all guard orders?
23. Do physical, functional, or other changes at the installation indicate the necessity for, or feasibility of, (1) establishing additional guard posts, or (2) discontinuing any existing posts or patrols?
24. Is two-way radio equipment installed on all guard patrol cars?
25. Are duties other than those related to security performed by guard personnel?
26. Are guard patrol cars equipped with spotlights?
27. Does each guard on patrol duty carry a flashlight?
28. Do guards record or report their presence at key points in the installation by means of (1) portable watch clocks, (2) central watch clock stations, (3) telephones, or (4) two-way radio equipment?
29. Are guard assignments and patrol routes varied at frequent intervals to obviate and established routine?

FIRE PREVENTION CHECKLIST

Security safety and protection go hand in hand with fire safety and protection. Recently a fire occurred in a warehouse that caused the company a financial loss of $3 million. If an assault on the building was committed by six thieves, the company losses would not have reach $3 million. The following checklist was prepared by the Boston Fire Prevention Division. It will aid you by pointing out fire vulnerabilities that, if corrected, may have an effect on your client's insurance policy and overall security.

Fire Vulnerabilities

Construction Type (Check)

1. Fireproof
2. Semi-Fireproof
3. Heavy Timber & Masonry
4. Light Wood & Masonry
5. Metal Frame
6. Wood Frame

Number of Stories ______________________________

Height ______________________________

Width ______________________________

Depth ______________________________

No. of guest Rooms or Occupants ______________________________

Certificate of Occupancy ______________________________

Certificate for Places of Assembly (List location and approved capacity)

____

____

____

____

____

Fire Department Licenses, Permits, or Flame-proofing Certificates (List) (Are They Up To Date?)

____

____

Other ______________________________

Maintenance Contracts (Name & Address of Contractor)

For cleaning range hoods, ducts, etc.

____

____

____

____

For Maintaining Fire Alarms System

____

____

____

____

Fire Alarm System

Type (local, etc.) ____________________________

Was it tested at this inspection? ______________

Did it operate properly? ____________________

Is there a municipal, automatic, ADT, etc., alarm box on the premises?

____

Is it connected to the Local Alarm System? __________

Are there any features to this system that should be noted? (Pre-alert, manual wound, etc.) ______________________

What means are provided for notifying Fire Department of fire? _____

____

____

Will a responsible person be available 24 hours a day to transmit an alarm? ____________________________

____

Emergency Lighting

Type of system ______

Are all units operational? ______

Are all areas covered by system? ______

Is system regularly inspected and tested as required by law? ______

Are records maintained of such tests? ______

Are fire drills held for employees? ______

Are competent watchmen assigned and on duty from 9:00 p.m. to 6:00 a.m. as required by Chapter 143? ______

(Two men required for 100 or more sleeping rooms in buildings 4 or more stories high, one man required for 50 or more sleeping rooms in buildings 3 stories high).

Sprinkler System

Is it complete or partial ______

Are there any areas that should be sprinklered? ______

______

______

______

What supervisory features are provided? ______

______

______

Are sprinkler valves open? ______

1. Gauges registering properly? ______
2. Sprinkler heads free and unobstructed? ______
3. Where does sprinkler alarm sound? ______

Standpipe and Hose System

1. Conditions of hose and nozzle? ______________________________

____

____

____

2. Location of first aid standpipe. ______________________________

____

____

____

3. Locations and Fire Department Standpipes ____________________

____

Fire Extinguishers

1. Sufficient in number? (One for every 1500 sq. ft. floor area)

____

____

2. Is the type provided, suitable for the hazards? _______________
3. Condition of extinguishers ______________________________

____

____

4. Have they been recharged as required? (yearly) ______________
5. Are they accessible? ______________________________

Exits

Are there two independent exits from each floor? _______________

Any obstructions to, or in, doors, stairs, landings, corridors? _________

____

Do exit doors open OUT? ______________________________

Fire doors kept closed? ____________________

(Remove any wedges or other obstructions) (Remove any wedges or other obstructions) ____________________

Is there any location where additional exits are necessary? __________

____

Exit and directional signs (natural or artificial lighting) __________

Are there barred windows? ____________________

Or locked doors? (Clarify if answer YES) ____________________

____

____

____

____

____

Are enclosed stairs or fire towers available? ____________________

1. Location of same ____________________
2. Self-closing doors? ____________________
3. Proper lighting __________ Type of emergency lighting __________

   ____
4. Hand rails ____________________

Are there any transoms over room doors? ____________________

Locations of same ____________________

Interior Finishes of Floors, Walls and Ceilings

1. Is any acoustical ceiling tile used? ____________________

   Location of use ____________________
2. Is any questionable wall covering used? (Wood paneling, tile, fabric, etc.) ____________________

   Location of use ____________________

3. What type of floor covering is used? ____________________

Note: Samples of above should be obtained if possible and submitted for testing. Burden of proof is on occupant.

General Check—Any Area

1. Accumulation of waste paper, rubbish, furniture, etc. Where?

____

____

____

____

____

____

2. Housekeeping deficiencies ____________________

____

____

____

3. Stairs, ramps, elevators, vent, dumbwaiter and other shafts
    a. Are they enclosed? ____________________
    b. Are doors self-closing? ____________________
    c. What type of interior covering? (Must be non-combustible) ____________________

    ____

    ____

4. All chutes, including *rubbish* and *laundry:*
    a. Do they have closing doors? ____________________
    b. Are there non-combustible linings? (Building Code requires 1 hour fire resistant rating) ____________________
    c. Is there an automatic sprinkler installation in the chute? ____________________

d. Is there a fire detection unit provided? ________________

e. Are venting facilities provided which extend through roof? ________________

Are they adequate? ________________

5. Are there instances of temporary wiring, extension cords, or the use of electrical equipment which should be corrected? ________________

____

____

6. Are covered metal containers provided for rubbish, oily waste and other materials? ________________

7. What methods are used to dispose of rubbish? ________________

____

____

8. Is rubbish removed daily? ________________

Note: Plastic baskets and barrels cannot be used for rubbish.

Maintenance Shops

	Violations Found	Corrections Made
1. Carpentry	________	________
2. Plumbing	________	________
3. Electrical	________	________
4. Upholstery	________	________
5. Paint	________	________

Incinerator

1. Location ________________
2. How cutoff from rest of building ________________
3. Is it an approved type? ________________

4. Do feed hoppers close tightly? ____________________
5. Has it a spark arrestor? ____________________
6. How is rubbish handled in transfer from chute? ____________________

____

Heating Equipment

1. Location ____________________
2. How cut off from rest of building ____________________
3. Are all openings thereto kept closed? ____________________
4. Type of fuel used ____________________
5. Where and how fuel stored? ____________________

____

____

____

6. Individual responsible for this equipment ____________________
7. What defects were noted? ____________________

____

____

____

Emergency Electrical Service

1. Type ____________________
2. Location ____________________
3. How and when placed in operation ____________________

____

Utility Gas Service

1. Location ______________________________
2. Size ______________________________
3. Shut-off location ______________________________

Electrical Service

1. Location ______________________________
2. Shut-off location ______________________________

Heavy Fuel Oil Shut-Off

1. Location ______________________________
2. Is it operable? ______________________________

Outside Means of Egress

1. Number and type ______________________________
2. Location ______________________________
3. Condition ______________________________
4. Do they open out? ______________________________

Outside Entrances to Basements

1. Location ______________________________
2. Do they open out? ______________________________

Steamer Connections

1. Location ______________________________
2. What do they supply? ______________________________
3. Are they accessible? ______________________________

Air Conditioning

1. Complete or partial ___
2. Location of intake ___
3. Shut-off control ___
4. Dampers (automatic where passing through walls, floor, etc.) ___
5. Refrigerant used ___

Inspector check following from Outside Building

Outside ___

Inside ___

Location of nearest city Fire Alarm Box No. & Location of hydrants in relation to building ___

City hydrants ___

Private Hydrants ___

Condition, width and grade of Streets ___

Drives or approaches to Hotels ___

Can apparatus approach for emergency evacuation, etc. (clarify) ___

Building areas that are accessible ___

Building areas not accessible ___

Factors or obstructions influencing or hampering accessibility for Department operations (Such as narrow drives, tunnels, grades, overhead wires, fences, walls, gates, parked cars, building set back, etc.)

____

____

____

Exposures to this building

1. Adjacent buildings ______________________________

   ____

2. Grass, rubbish, brush, etc. ______________________

   ____

3. Combustible sheds, other ________________________

   ____

Buildings that this building might expose ______________

____

____

Remarks: ______________________________________

____

____

____

Recommendations: ________________________________

____

____

____

Note: Briefly describe the existing stairways including the number, type enclosure, their termination point, etc. ______________

____

____

LOSS PREVENTION & RISK ANALYSIS MEASURES

Risk Analysis: Theory of Vulnerability Reduction

The subject of this chapter relates to the implementation of operational applications in risk analysis (RA). The goal is to provide the reader with basic procedural guidelines, in order to facilitate a more viable and practical approach to contemporary vulnerability-reduction techniques. Accordingly, it is believed that increased chances of achieving successful operational results will emerge from these applications.

THE THEORY

The theory itself is straightforward in the sense that it is relatively easy to enumerate the steps required in the context of risk analysis. For it is noted that these steps do not differ significantly from those directed to the scientific method in general. In another sense, however, the theory is rather difficult to extrapolate. The real problem lies in providing an efficient vehicle that will aid the professional in the identification of the steps used or required in the planning of the successful security operation. In other words, the difficulty is not in identifying the steps of analysis, but rather in aiding the practitioner in translating these steps into actions appropriate to the specific problem area. Consequently, the emphasis of this paper is not on describing the specific steps of risk analysis but rather on bringing these already generally recognized methods to a practical application through a selected examination.

In a discussion of security-planning processes, the primary step of "writing down the problem areas" is obvious. This ensures a positive approach to contingency planning and provides the planner with a total overview of operational considerations. The reasons for this basic, but often overlooked, approach are several.

Primarily, this procedure ensures that the RA surveyor does not make what may be referred to as a "problem-definition oversight—" namely, solving the wrong problem right. Consequently, it also ensures that the right problem is not addressed inappropriately. Problem definition facilitates a valid determination of the contingency-evaluation process whereby any operational considerations are valid representations of the decision problems being studied, that they are factually significant, that the inquiry is completed within operational time limits, and that the final operation is effectively and efficiently accomplished. In short, that the RA process effected a successful security operation.

One additional observation should be noted. The procedures and guidelines provided in this paper may not be precisely appropriate to each individual inquiry. For each security operation will be different, and will require initiative, intuition, and imagination on the part of the surveyor to successfully define and accomplish the specific operational objectives. Every security problem is unique and must be independently evaluated. In this regard, it is imperative for the surveyor to understand that no one solution nor approach can meet every contingency. Nonetheless, the general principles and guidelines regarding the conduct of the RA inquiry may be easily incorporated into the problem-solving process.

This paper, for the sake of clarity, is organized in sections corresponding to the major steps generally undertaken during the risk analysis inquiry process. The steps include the following:

1. Survey Objectives,
2. Overview of the Security Plan,
3. Problem Evaluation,
4. Problem Development,
5. Analysis of Variables,
6. Program Validation, and
7. Implementation.

SURVEY OBJECTIVES

The primary approach to any RA inquiry is to determine what the surveyor expects to achieve from the operation, in other words, what the operation expects to accomplish. Generally, with respect to the security industry, this involves either the safeguarding of a valuable commodity or the safe secure transfer of that commodity from one point to another.

The definition of objectives should be in terms of the goal to which the inquiry is directed. The surveyor should take special care to ensure that the goal is not defined in too narrow a context. Conversely, the objectives of the inquiry should not be needlessly broadened in an attempt to solve all the problems of security, generally, in one comprehensive and usually unsuccessful approach.

Thus, to avoid the risk of not adequately understanding or defining the objectives of an RA inquiry, the surveyor is usually instructed to devote effort to problem definition. This is valid advice, and should be considered. However, the real issue may be a different one. The question of devoting effort to develop a continuing dialogue with the individual or client who commissioned the RA inquiry must be addressed. The surveyor must understand the specific operational needs of the client. For unless the continued communication is maintained between the surveyor and the client, the surveyor will not be able to establish the perimeters of the survey. Moreover, a lack of understanding of what the client actually wants to be protected against is unavoidable.

The key, therefore, is how the surveyor must conduct himself during communication with the client. Further, how a sequence of discussions may provide a thorough understanding of the client's problems and needs. In this area five considerations should be noted:

1. The surveyor should both listen to and hear what the client advises. The obviousness of this is exceeded only by the frequency with which it is not done.

2. The surveyor must place himself in the client's position, and view the problem from that perspective. He must also try to gain insight into the client's personal characteristics, which may involve certain biases.

3. The surveyor should also understand the organization that the inquiry will influence. He should be thoroughly familiar with its reporting relations, functions of various groups, and prior inquiries of a related nature.

4. The surveyor should refrain from pushing his own preconceived notions of the problem, or implicitly adjusting the problem to fit prior conceptions of the approach he would like to adopt.

5. Most important, the surveyor should not let statements or comments of the client pass unexamined. Too often the RA surveyor sits and out of deference passively listens to the client's interpretations

of the problem areas. This is counterproductive, for a continued exchange of questions and answers between the surveyor and the client is necessary for effective problem definition.

Gaining an understanding of the objectives of the inquiry before it begins, unfortunately, may not be sufficient. It is necessary to anticipate and be aware of how program objectives may change with time, or with the level of management that becomes involved with the inquiry. In this instance, the problem may be easily rectified, once again, through continued communication with the client. For not only will the client feel an active participation in the total security project; but it also ensures the surveyor of no unanticipated surprises at the time the completed risk analysis is tendered to the client.

OVERVIEW OF THE SECURITY PLAN

The second step of an RA inquiry is the development of a security plan. The process of performing an inquiry is often a creative one. As a result, the question is raised of how to establish the planning process. The answer is clear if one considers planning as a creation of deadlines, that is, of points in time at which certain accomplishments are to be completed, rather than as the specification of the particular way in which the work is to be accomplished. The former is a necessary control, only the latter is creative. The act of establishing deadlines, however, is in itself a stimulus to creativity and accomplishment. Many are familiar with the last one or two days before a deadline in which all the work gets done. Thus, adherence to time deadlines (created by plan) forces creativity, leads the surveyor to conclusions, and in general contributes to the success of a project. Conversely, delays and missed deadlines lead to indifferent performance on the part of the surveyor and may effect disinterest on the part of the client.

Failure to meet schedules frequently occasions additional difficulties. Clients are often responsive to problems of immediacy, especially in the area of security. If the conduct of the RA inquiry is unduly long, the problem, or at least the client's perception of it, changes. The result is that the surveyor may solve a problem other than that which is of current concern. To be sure the client may be at fault; but the surveyor may have also contributed to the difficulty. He may not have properly planned where he should be at each point during the problem-solving process, disciplined his conduct, or adhered to the required schedule.

However, extreme care should be exercised in the development of the security plan. Although the evolution of a work plan is useful to discipline the activities of the surveyor and enables him to proceed on schedule, it may prove less effective in coping with client avidity, which is sometimes a characteristic of successful RA inquiries. Client avidity usually results from the client's recognition that the inquiry, or its recommendations, may be potentially more useful to him than previously thought. In this instance, the work of the surveyor is usually expanded to the point of the surveyor being forced to solve a number of different security problems that were not included in the original perimeters of the inquiry. Further, it may also prematurely involve the surveyor in the implementation of operational guidelines that were not effected through sound planning processes. In that same context, premature expansion of a potentially successful inquiry usually precludes the successful completion of the original project. More often than not this premature expansion of the inquiry delays schedules, and usually the surveyor is unable to complete the project within the previously established time frame. Operationally, a security plan is simply a time schedule for the accomplishment of the inquiry steps, which may be further divided into specific tasks. For example, the steps concerned with the development of an approach to a cargo-control problem may be composed of the following individual tasks.

1. For each identified problem area, develop the perceived solution method.
2. Describe and document the solution method. The resulting report will describe the inputs, outputs, and control steps necessary to obtain a solution to each problem. The control steps must be specifically expressed and verbally explained to the client.
3. Test the proposed solution methods on small-size selected sample problems. This step will ensure the feasibility of the methods recommended for the solution.
4. Revise the solution techniques as indicated by the selected sample application.
5. Prepare a summary of the sample results and add this to the completed documentation report.
6. Prepare and deliver presentation on the results of this method of vulnerability reduction.

The development of specific project planning (the detail just illustrated) is often essential to the determination regarding valid estimates of time and resources necessary to complete a project. Factoring the total inquiry into its constituent steps is possibly the best way to ensure valid estimates of the required time, budget, and staff before a project begins. In other words, the project plan increases the probability that the inquiry can be completed on schedule and within the required budget constraints.

There are two additional considerations in developing the project plan that should be briefly mentioned. The first of these is the importance of providing for project deadlines. As mentioned earlier, these act to discipline the surveyor to remain on schedule. Deadlines, useful in this respect, include submission of documentation, presentations to clients, and the formal initiation of new approaches to project steps.

The second consideration is the allocation of resources for the analysis of information collected during the inquiry, for purposes that might be unrelated to the inquiry's direct concerns. There are three reasons why this activity is important.

1. The RA surveyor, merely because he is collecting and thinking about an activity, may uncover understandings of more benefit than those associated with the original inquiry. In other words, there is always a chance of an association with another security-problem area that may be benefited, indirectly, from avocations recommended in the initial inquiry.
2. Throughout the problem-solving process, it is the responsibility of the RA surveyor to report to management and others involved with the project on the progress of the inquiry. Clients, however, are usually results oriented, and during a long project their interest may dissipate, especially if the status reports are concerned only with progress and development rather than what is to be implemented. Often, a straightforward verbal analysis of information collected is sufficient to maintain the client's interest and support.
3. Finally, such analysis and feedback can aid the information collection activity. As previously indicated, increased client participation in the RA inquiry usually benefits the surveyor. In addition, client employees will be well aware of the increased activity in the

security area, thus making the employees less likely to participate in unauthorized enterprises.

PROBLEM EVALUATION

The next step, that of problem evaluation, is of fundamental importance in the conduct of a risk-analysis inquiry. It marks a significant fork in the road, with one direction leading toward failure and often frustration on the part of the surveyor. It involves additional discussions with the client, often the first contacts with the actual line and staff personnel of the client, and the collection of additional information necessary to gain a thorough understanding of what the specific problem area is, what has occurred in the past, and what is likely to occur in the future. The conclusion of this specific task area provides the surveyor with a framework for the recommended approaches to the problem areas being studied. And establishes the direction of all future work in this area. Its importance warrants a careful discussion.

The first issue of problem evaluation is often to decide whether or not the total problem should be broken down into smaller separate problems that may be studied individually, in parallel, or sequentially. Doing this is frequently necessary fro dealing with large and complicated areas of activity. The process of breaking down a large problem in its logical subparts, does, of course, lead to some optimization. Its effect may be minimal, however, if the problem is divided at points of the weakest interaction.

To illustrate how a surveyor proceeds in this task, let us use a hypothetical example of a production and distribution planning problem and see how it is divided into meaningful, but yet solvable, subparts. The problem is to develop inventory-control procedures on a weekly basis over a four-week planning period. During the initial problem formation stage, several questions are identified as important. Some of these are as follows: (1) With only a four-week planning period, how can the need for seasonable buildups of inventory be anticipated? Or in another way, what should be the desired level of inventory control at the end of a four-week planning period, with an adjustment for seasonal buildups? (2) What should be the optimal controllable warehouse inventories? (3) How should plant shipments to warehouses be spaced so as to take advantage of the best methods of inventory control?

To deal with all these issues, yet to wind up with a practical approach to problem resolution, indicates that the problem should be subdivided into four principal components: (1) Consideration should be given to the generation of forecast demands for inventory on a weekly basis over the four-week planning period. (2) Given the demand forecasts, the surveyor would determine the optimal inventory-control levels at the warehouses for the end of the four-week planning period. From these, and based on a knowledge of current inventory levels, minimum control requirements can be determined. (3) The next step would use annual sales forecasts, provided by the client, to set inventory control and procedures for each four-week planning period during the year. The resulting inventory control approaches would provide guidelines for total inventory and at the same time, reflect the shorter-period reduction considerations. (4) Finally, the surveyor would utilize the inventory requirement provided by the client, and the optimum warehouse inventory-control levels, to set general security guidelines for each of the four-week periods.

In developing an overall approach to the type of problem illustrated above, the surveyor should generally proceed as follows.

1. Start with the central decision problem to which the inquiry is addressed. In the example above, this is the four-week planning period.
2. Ask what inputs and outputs are required of the procedures for the central problem. This will aid the surveyor in determining the scope of the sub-problems which may need to be addressed.
3. Determine if there are already existing means of obtaining the required inventory-control measures. If so, these may be used. If not, additional procedures must be developed.

Following an analysis process of the type outlined above will permit the surveyor to define manageable problems, better integrate his efforts with current practices or procedures, and increase the chances of successful implementation of recommended programs. At its conclusion, the surveyor should know what the programs he plans to develop are to accomplish, and how they are to be sued in the context of the client's problems with which he is dealing.

Some thoughts should also be devoted, even at this early point, to proper procedural maintenance. Maintenance is essential for the long-term success of a risk-analysis inquiry involving the implementation of new procedures and guidelines. If this consideration is left until implementation, chances are that maintenance methods will be inadequate, and the chances for long-term procedural successes will substantially decrease. The following checklist may help to ensure successful maintenance design.

1. Will there be a means for the continual updating of security procedures?
2. Will the documentation be adequate to ensure that others can approve and update the procedures?
3. Has provision been made for the operation of the procedures within the routine activities of the client's firm?
4. Can the procedures be easily modified to reflect changes in the actions and routines of the client's activities?
5. Has adequate consideration been given to determining ways of using the recommended to provide information on possible weaknesses and errors in the recommended system?

A second issue that must be resolved during the problem-formation phase is to determine the extent with which the programs are to be developed. Consideration of the budget available, the project schedule, and the objectives of the inquiry should be the principal factors that influence this decision. Often in making choices in such matters, the surveyor mistakenly associates the extent in which he develops a program with its validity. *The extent is an incomplete measure of validity.* Most important, the surveyor should ensure that he develops what may be referred to as "unbiased" programs and procedures. Put another way, any analysis is subject to error. These errors can sometimes be overestimates, or sometimes underestimates. On the average the procedures should provide "fair" predictions of real-world results. In this regard, the extent to which the program developed is therefore not necessarily a significant influence of its unbiasedness, but rather a "judgment call" on the part of the surveyor.

Once the total problem has been factored into appropriate sub-parts and an idea of the desired degree of extent obtained, the problem formation may continue. The next phase of problem formation involves identifying the scope and dimensions of the programs or procedures to be developed. For this, five specific tasks of problem evaluation are noted.

Task 1. The Determination of Problem Dimensions

This first task of the problem-evaluation phase may be thought of as the determination of the variables of the problem in an analytical statement. For example, specification of the problem dimensions may include the determination or classifications for a number of security personnel, logistical support, definition of the planning period, or the time intervals as they relate to the total security project.

Task 2. The Determination of Client-Controllable Decision Variables

This second step involves the determining of variables that are under the control of the client. They need to be explicitly determined during problem formulation in order to distinguish between the "unchangeable facts" of the problem and what the client is willing to consider as changeable. In other words, what are inputs to the planning problem as opposed to the alternatives or variables of the decision? In the restructuring of a security force the decision variables may be related to the number of personnel to be employed and perhaps the rules governing "non-security" personnel conduct. Whether or not to use an in-house contractor for security services may not be a decision variable of the problem. Of course, if it is not, this should be the result of an explicit analysis and decision that involves both the surveyor and the client.

Task 3. The Determination of Uncontrollable Variables

A third task of problem formulation is to determine the uncontrollable variables. These are the aspects of the problem that are not controllable by the client but that may have an impact upon the operation of the activity on an intermittent basis. In the evaluation of a security problem, these may include, but not necessarily be limited to, labor disputes, sabotage, natural disasters, or simply morale problems within the security organization itself. In general, failure to include such contingencies in recommended procedures may lead to problems that are not valid in their representation, and even worse, if adopted, that lead to inappropriate and misleading policies.

Task 4. The Determination of Technology

The determination of technology is the fourth task of problem formation. The technology is described by the constraints and perimeters that define the limits of the inquiry itself. In any inquiry addressing vulnerability-reduction techniques, the question of sophisticated hardware must be evaluated. Perimeter controls, barriers, electronic security systems, and lighting systems are key elements in a total security package. The surveyor must possess in-depth knowledge of the various systems available, and be able to provide the client with a contemporary evaluation of the current systems and their applications.

Task 5. The Determination of Measures of Effectiveness

The measures of effectiveness are the criteria used to judge the merits of particular solutions to the problems being evaluated. Risk-analysis inquiries should involve several measures of effectiveness, the most important being those that come to the surface during the definition of objectives. Often the choice of proper procedures is not obvious except in retrospect. In the case of the proposed vulnerability-reduction programs, the positive results will not usually be readily apparent. In most cases an effective program will not manifest itself until the client reviews the appropriate records and personally observes a lessening of employee thefts, an increase in employee productivity, or perhaps a decrease in the number of security-related incidents.

PROGRAM DEVELOPMENT

The fifth step in risk analysis is concerned with program development. The program expresses the interaction among the controllable decision variables, the uncontrollable variables, the technology, and the measures of effectiveness. Its proper development is a key element for project success. If done improperly, the program can be misleading in its inferences, or uneconomical in its use. If done properly, the program can contribute substantially to the client's proper performance and the organization's operating efficiency. Valid program development requires that substantial amounts of information be collected and carefully analyzed to ensure that a proper understanding of the problem area has been achieved.

A first task of program development is the resolution of whether or not the required numbers or relationships are feasible to develop. The validity of the inquiry hinges on this question in the case of most security-related problems. On the other hand, it is also reasonable to measure the increased incidence of security-related problems should a shift of guards be deleted.

But an inquiry requiring too many unknowns makes it extremely difficult to construct relationships among variables. There is, unfortunately, no valid answer as to how much is too much. The desirability of undertaking an inquiry when little is known about what the client's actual needs are, depends to a large extent on how decisions are currently being made in the organization. If the surveyor is involved with several staff and managers, each of whom contributes subjective judgments, and the decision process is one of consensus, it will be extremely difficult for the RA inquiry to improve upon results. On the other hand, if the decision process involves but one person with limited input information, a program might nevertheless lead to improved decision making.

Even if there is a lack of understanding of critical parameters and relationships, a program can break the problem into smaller elements each of which is more easily handled. It may also provide a discipline to the client, requiring him to use at least what information is available in the best possible manner. At the conclusion of the program-development step, a precise statement of the specific problem should be made available to the client. Understanding of the problem should be sharper, and unclear or uncertain relationships identified. Only after the surveyor and the client agree on the specific problem is the surveyor ready to develop a valid approach to problem resolution.

ANALYSIS OF VARIABLES

Simultaneously with the program development step, it is necessary to determine how the problem is to be solved specifically. This involves several stages in itself. For example, should a standardized-manual approach be adopted; should the program recognize the randomness of variables or should a deterministic approach be adopted; should certain client interpretations be recognized or are personal observations sufficient; can "state-of-the-art" solution methods be used or must new methods be developed? In short, what assumptions need to be made and methods developed in order to make the program practical?

This step may also involve the "testing" of proposed problem resolutions or approaches to ensure, at least for small test problems, the feasibility and the validity of the methods being presented. All this, of course clearly indicates the interdependence of problem formation, program development, and solution. In any event, for those programs for which it is necessary to develop a broad-based application, the surveyor should start with the best possible formulation of the problem, even if that formulation is clearly unsolvable. From this point, the surveyor can then begin to make simplifying assumptions as needed. The advantages of this approach are several. First, there will be a realistic problem formulation. The surveyor is thus assured of a proper understanding of the problem. Had he never developed and oversimplified the problem, he may never understand the problem with which he is dealing. Second, the surveyor is better able to understand the implication of any necessary simplifying assumption, in terms both of the realism of the program and of the impact of his assumptions on the final program recommendations. Finally, if it is necessary to develop an abstract method for the problem solution, the surveyor is able to start with what would be a functional procedure and depart from this. Having such a frame of reference will aid the surveyor's chances for developing a practical, efficient program from an abstract theoretical method.

In most instances, the surveyor, with respect to this relationship with the client, is usually given the choice between obtaining:

1. An optimal solution to a simplified version of the problem, or
2. An approximate solution to an exact formulation of the problem.

Experience strongly suggests the desirability of the second choice. In part, this is due to the lack of confidence in an inquiry that may evolve from a failure to recognize a significant aspect of a specific problem area. Nothing can ruin a relationship between a client and a surveyor faster than a recommendation that is clearly wrong because the program did not take into cognizance the characteristics of a problem that the client believes to be important. In practice, one decision will not solve all the security problems for a particular client; but the many factors that interplay and interact on a decision problem are a fact. The client may need help in considering these; therefore, a complete representation of important variables, even if the proposed solutions are general in nature, will usually be of more value for a client than a complete solution to a problem that the client feels is not critical.

PROGRAM VALIDATION

It is during this step that the information needed for validity testing and program implementation is collected and evaluated. The importance of this step cannot be overemphasized. It is imperative that a surveyor validate a program prior to implementation. Without a systematic effort directed to contingency validation, any risk-analysis inquiry, regardless of its complexity, is doomed to failure from the onset. In this context, the proposed methods of vulnerability reduction should be examined for:

1. consistency-to ensure that results are logical when the major parameters of the inquiry, as prescribed by the client, are varied and broad based;
2. sensitivity-to ensure that the relative magnitude of the proposed changes in client operations are also in concert with smaller, peripheral operations within the client's organization;
3. plausibility-to ensure that the inquiry recommendations actually respond to the special situations within the framework of the client's operation;
4. workability-to ensure that the recommended procedures that are to be used by the client in support of management decisions have the capacity to offer response in the time frames appropriate for these decisions.
5. flexibility-to ensure that the procedures are easily adaptable to slight alterations in the client's plans or operations.

If for some unforeseen reason the proposed programs and procedures do not meet these criteria, the surveyor may, out of necessity, have to redesign areas of the program's application.

IMPLEMENTATION

In some cases the RA surveyor may believe that his responsibility ends with the development of a program. This is not the case. In actuality, the real difficulties with an inquiry often surface only during this final project step, that of implementation and maintenance. The effectiveness of the total program will depend heavily on obtaining the cooperation and participation of the client in the decision problem being studied or the operation being affected. Much of the resistance that may be encountered can be attributed to nothing more serious than a lack of understanding and a reluctance to change. These may be overcome, but it is a responsibility of the RA surveyor to devote the time and patience required.

S successful security-program implementation involves a preparation of training materials, the development and presentation of orientation seminars, and the use of realistic case-study applications. Although many clients, by the time an inquiry is completed, may be familiar with the procedures developed, it is still important to ensure that their understanding is accurate, and that suitable materials for training others are available in order to maximize the program's long-term usefulness. It is only when the surveyor is willing to cooperate and work in harmony with those whose activity he is studying, that the proper partnerships can be formed and the risk-analysis inquiry successfully concluded. Implementation is a process of "hand-holding"; without it the surveyor and the client do not establish the needed communication and cooperation to implement the procedures.

Another important, but often neglected, aspect of an inquiry that involves the implementation of new procedures is the short-term problems and disruptions it causes. These are often inevitable regardless of the extent to which the procedures are validated. Care should be taken to ensure beforehand that all interested parties understand fully the high likelihood of difficulties that will be incurred in the operation of a new security program. More important, however, is the need for the surveyor to have available:

1. a carefully worked-out plan directed to the resolution of implementation difficulties;
2. the resources necessary to make, on a timely basis, the changes required to correct difficulties that may arise; and
3. the client's approval to subtly alter the program to resolve the anticipated and apparent difficulties encountered.

CONCLUSION

Those involved and affected by the implementation of new security programs and procedures must recognize the need for a period of time in order to "iron out the bugs" always found in a new system. As previously indicated, the theory of risk analysis and vulnerability reduction is straightforward in principle; however, translating the theory into practical applications is often a prime difficulty encountered by the surveyor. As in all human endeavors, there are no absolutes or guarantees. In that regard, the manner in which a RA inquiry is conducted is perhaps the key to successful resolutions of specific security problems. The process of vulnerability reduction is a calculated, methodical enterprise. There are no shortcuts, no magic answers. But if the steps previously indicated are applied to this process, a valid security program, regardless of complexity, may be adapted to virtually any conceivable situation.

LOSS PREVENTION SUGGESTIONS

1. When a trucking terminal has guard service only from midnight Friday to midnight Sunday, it should insist that the guard service take a complete inventory of the yard when coming on duty and again when ending duty. This should include tractor numbers, trailer numbers including seal numbers, and straight-job and seal numbers. Each inventory should be accomplished on a separate piece of paper, placed in an envelope, sealed, and pushed under the facility manager's office door. There will be no question as to placing responsibility if a tractor, trailer, or straight job is missing or has been opened.

2. When physically checking a social security card against an application, pay close attention to the two middle numbers. If the last digit is odd, such as 1, 3, 5, etc., check closely, for it may be fictitious. Mainly aliens and individuals from prior to 1930 have odd numbers here.

3. An effective way to prevent trailer swing-type doors from being opened within a fenced yard is to park two trailers back to back if there is enough room in the yard. If not, erect highway-type guard rails approximately six feet high and back the trailers close to the guard rail.

4. If a company desires to fingerprint, photograph, and have new employment applications filled out by drivers and dock workers and there is union objection, in some states the Alcohol Control Board has a requirement for all three procedures in terminals

where liquor is handled. This requirement is seldom enforced, but it can be used as a reason to accomplish the desired results.

5. When a company desires to place an undercover person on a terminal doc, be sure to consider a company porter, for he would have reason to be anywhere in the terminal.

6. When a trucking terminal closes down over a weekend, consider locking off the cock area completely. This can be accomplished by affixing locks to all dock doors and to any doors leading from outside areas onto the dock. Apply a hasp so that when the last management person leaves on Friday night or Saturday morning, that person locks the door and places a seal through the hasp and lock. He then records the seal number. The first management person to return to the dock removes the lock and records the seal number, which can be checked against the seal number from the previous Friday.

7. Freight that sits on a dock for several days is the freight most likely to be stolen. One way to know the amount of time freight has been on a dock is to date it each morning after the trucks have left the terminal. Any freight that receives a third date should be recorded and the list given to the terminal manager. Using chalk of a different color for each date is very effective in dating the freight.

A MANAGEMENT CHECKLIST FOR THE MOVEMENT OF CARGO

There are hundreds of management-oriented theft-prevention methods that can be built into the processes by which cargo is shipped, transported, received, stored, and documented. Of necessity, therefore, the several dozen preventive steps that follow do not represent an all-embracing checklist but are important as illustrations of the *type of thinking* that executives can and should bring to bear on the cargo-theft problem. The suggestions below are hardly revolutionary nor revealed for the first time in this publication. Procedural in nature, they do not require hardware to implement, just the will to do so. Though presented in piecemeal fashion here, such measures should not be applied on an ad hoc basis but molded to fit local conditions and fielded as a well-coordinated team. To facilitate future reference, the numbers preceding those measures or approaches considered particularly applicable might be circled.

1. Consignee management should instruct its receiving department to notify purchasing when incoming items arrive. This will help prevent fraudulent purchase orders originated by someone outside of purchasing from getting into the flow, and will force receiving to make a careful count.
2. To help assure timely detection of thefts occurring before goods get into the consignee's record system, request purchasing personnel to contact the supplier directly when an order is not filled within a reasonable time.
3. To prevent spurious purchase orders and subsequent thefts, prohibit purchasing department from receiving ordered merchandise and from having access to such merchandise. Likewise, assure that receiving personnel do not perform purchasing duties.
4. Only specified individuals should be authorized to check in merchandise received. Unless responsibility is fixed, shortages can be blamed on others easily.
5. Consignee employees who check incoming goods should reconcile such goods with a purchase order and remove goods to the storage area immediately thereafter. Absence of a purchase order could mean that the merchandise was ordered fraudulently with the intention of removing it before it got into the record flow. Prompt transfer of goods to storage not only gets them into the records flow but also removes them from a traditionally high-theft area.
6. Receiving personnel should use a pre-numbered form on which to record delivered merchandise, and copies should be sent to purchasing and accounts payable. This will help deter destruction of receiving records and theft of merchandise. Failure to furnish purchasing with a record will spur an investigation, and failure to advise accounts payable will result in a complaint by the supplier.
7. Regarding outbound shipments, freight bills are totaled and compared to the number recorded on the manifest discrepancies are promptly corrected. The same holds for incoming shipments except that discrepancies are immediately reported to the terminal manager and/or security director of investigation.
8. Freight received without accompanying documentation should be stored in a secure place.

9. Record the number of shipping documents given to shippers or loaders. When the documents are returned, count them again and compare totals.
10. When returned by the local driver, delivery receipts should be compared with terminal control copies and all bills accounted for.
11. Consignees should not delay taking delivery of goods. Anticipate difficulties regarding import license, exchange control, or other regulations. Those who have taken advantage of free time in customs and of free storage time at carrier terminals often find that the practice is penny-wise but pound-foolish. For example, an importer of canned goods took prompt delivery and suffered only limited pilferage in contrast to the heavy losses of his procrastinating competitors.
12. Require positive identification from pickup drivers to ensure that they are legal representatives of the carrier. Record license numbers, especially on rental vehicles.
13. Prepare *legible* bills of lading and other shipping documents, which are manufactured from a paper stock that will hold up under multiple handlings. Try to use classification descriptions instead of trade names, and avoid listing values.
14. Periodically rotate drivers among runs. Otherwise, there is too great a chance that they might develop contacts for collusion. Beware of drivers who request certain routes despite the lower wages associated with those routes.
15. Change truck stops frequently.
16. Develop incentive plans to control losses-payments to employees being based on reductions in insurance premiums and/or actual losses.
17. On multi-piece shipments, shippers should label each package. As the driver instruction manual of one carrier reads, "The driver must check all shipments to determine that each piece is legibly, durably, and properly marked. The name and address of the shipper must be shown on each piece of freight in any shipment. The marking on each article should be checked to determine if the consignee's name and address is the same as shown on the airbill. Drivers must be certain that the marking will not tear off when the shipment is in transit."

18. Exposure to loss often increased with higher turnover of personnel on shipping and receiving docks.
19. Advertise your security efforts in high-theft locations.
20. Know employees on *all* shifts.
21. Do not advertise on trucks. For example, "Smoke Brand X-distributed by . . . "
22. Run radio and TV spots indicating the convenience and other advantages (such as service) of buying through regular channels.
23. Request truck rental companies to post signs warning users that the rental agent is cooperating in theft prevention.
24. Conceal or seal in a pouch the papers covering a load.
25. Provide cargo checkers with self-inking identification stamps. When receipting for cargo, in addition to affixing his signature on the receipt, the checker stamps the document, thereby clearly identifying himself.
26. Utilize color-coded vehicle passes (keyed to specific areas in the terminal) and time stamp them.
27. Establish advance-notice procedures whereby consignee is notified at least 25 hours prior to the arrival of sensitive shipments. Alert intermediate points as well.
28. Negotiate with carriers for what one large shipper calls "signature security service" for certain kinds of shipments, whereby a signature and tally are required from each person handling the shipment at each stage of its transit, from point of origin to destination.
29. Analyze claims to determine type of cargo most subject to theft and where it is being lost.
30. Periodically review carrier performance to identify those having a high incidence of loss or damage.
31. Receiving stations should assign a trusted employee to review advance manifests or, if none, the documents arriving with the cargo to identify and segregate for special attention theft-prone cargo. Such "paper alerts" should also be supplemented by actual examination of the cargo. For example, a manifest

described one shipment as "electrical equipment," but the carton identified the goods as calculators. Relying on the manifest, a cargo handler treated the shipment as general cargo; one of the calculators was later stolen.

32. Security rooms for valuable cargo should have adequate inventory-control procedures. In one case, a bale of furs was lost for two weeks because of a lack of such controls.
33. Maintain a tally as cargo is transferred from vessel to terminal. This is frequently, if not usually, omitted, with the result that (a) detection of thefts are delayed until consignee complains, (b) buck passing is promoted inasmuch as no one knows whether the goods were stolen before or after the ship was unladen.
34. Remove loose or broached cargo to cooperage shops as soon as possible.
35. Use shipper's initials rather than full name on labels if the full name would tip off thieves to the nature of the carton's contents.
36. Seal cartons with a tape having a special design and color. If a carton is opened and resealed, this would be clearly evident.
37. Indicate lot numbers on each carton and on the bill of lading in order to highlight a short carton.
38. As practicable, insist on piece counts when cargo is moved to and from vehicles and in and out of storage areas, vessels, railcars, aircraft, etc. And insist on clear identification of those who conduct such counts—driver, checker, receiving personnel, terminal cargo handler, or whoever. The tow parties involved in a cargo transfer should not take one another's word regarding the count. Accountability then becomes blurred. As a carrier executive advised, employees who check cargo must be told, "You are individually responsible. You must know. You must count." Among his instructions to drivers: (a) "If the bill calls for 'CS. No. 1234,' don't accept a case marked '4567' for it." (b) "A driver should never accept a shipment described as 'one bundle tires.' The airbill should indicate how many articles are in the bundle. For example 'one bundle (3) tires.'" (c) If a shortage exists in a shipment, determine the exact piece short. "If it is shoes, the exception should be '1 cs. Shoes short.' A general statement such as '1 cs. Short' is not sufficient…."

39. Load each delivery vehicle in conformance with its routing so that unloading delays-with concomitant unnecessary exposure to theft-will be minimized.

40. Segregate shipping from receiving areas, inbound from outbound cargo.

41. Require company name of carriers to be shown on all equipment. Do not accept temporary placards or cardboard signs as proper identification. Nor should lease agreements for rented vehicles be considered as sufficient identification. Independent verification is necessary.

42. Release cargo only to the carrier specified in the delivery order unless a release authorizing delivery to another carrier, signed by the original carrier, is presented and verified. Accept only original copies of the delivery or pickup orders.

43. Prelogged delivery or pickup orders should be safeguarded from theft or unauthorized observation. Verify identity of carrier and carrier employee before releasing a prelogged pickup order. An operator of large terminals notes the potential advantages of prelogging: "We encourage truckers to bring their documentation to the terminal the day before they deliver cargo. We prepare our receiving documents from the truckers' papers and when trucks arrive, give priority in handling to the loads for which we were furnished advance documentation. Cargo handling is expedited, checking is more precise, the documents themselves are more accurate. Our cargo accounting has improved significantly...." This confirms the observation in a carrier task-force report: "Reforms in paperwork to eliminate bottlenecks and to raise accuracy also make it less easy to smudge the responsibility for cargo and cargo records."

44. Restrict access to cargo documentation to a need-to-know basis. Systems assuring strict accountability for documentation are as important as those designed for the cargo itself. For example, after several thefts in a terminal involved stolen documentation as well as its cargo, an "internal release order" was devised. The cargo handler who is to retrieve a shipment in the terminal is given the release order, which describes the cargo and its location. Source documents remain in the custody of the clerk preparing the release order. The clerk retains a copy of the release,

on which he records time of preparation and the name of the cargo handler, who brings the shipment to his control supervisor, who, in turn, verifies the identity of the cargo handler and the description and quantity of cargo to be delivered. The supervisor requests the signature of the trucker, after recording date and time of release. Finally, the release order is returned to the clerk who prepared it.

45. Release seals to as few people as possible. Require all persons handling seals to maintain tight control over them.
46. Log movements of containers into or out of a holding area. Indicate date, time, seal number, truckman and company making pickup, and registration number of equipment used.
47. Accept only legible documentation.
48. Obtain clearly written signatures-not blurred initials-on cargo documentation.
49. Neither accept nor prepare penciled documentation.
50. Cargo entering a terminal from shippers or other stations should be thoroughly inspected, accurately counted, properly classified, and immediately stored. Paperwork should reflect all decisions and actions taken.
51. The employee withdrawing goods from storage should be different from the one actually releasing the merchandise.
52. An integral part of terminal security is a workable, accurate cargo-location system. Delays in, or confusion over, removing cargo from storage increases the risk of theft or pilferage. Among other things, a good locator system does not give cargo handlers the excuse to wander all over the terminal when looking for a shipment.
53. Devise procedures to minimize terminal congestion and poor housekeeping, which result in obstructed visibility of cargo, misplaced cargo, less efficient checking and handling, and other situations promoting theft and pilferage. States a highly knowledgeable source interviewed for this publication, "The real enemy of security is congestion. When goods pile up, you lose control, no matter what procedures are in effect." Many carriers try to combat this by discouraging consignees from delaying

pickup or acceptance of cargo. In the event that strikes hit other modes or carriers, some carrier terminals have readied emergency plans by which to handle in an orderly fashion the anticipated extra flow of cargo, such as through a pickup and delivery appointment system for shippers and consignees.

54. Establish space standards, which fix the quantity of cargo that can be safely stored in a given area or terminal.

55. High-value air freight should be the last to be loaded at origin and the first to be unloaded (and piece-counted) at destination. If practicable during surface transport, "bury" high-value shipments in the other cargo so that an unauthorized opening of trailer or container doors will not immediately expose the valuable articles.

56. To facilitate timely detection of theft, among other reasons, a weekly or twice-weekly telephone conference should be arranged between a carrier's various terminals (or a consignee's various branch stores) to reconcile cargo overages and shortages experienced at different locations.

57. Shippers should await the arrival of the carrier before loads staged for shipment are final-counted. Otherwise, between the time of the count and the arrival of the carrier, pieces could be added surreptitiously, and then hauled away through collusion between an inside employee and the carrier's driver.

58. Shippers should not permit customers to dictate routing. If every customer were permitted to route his orders, dock congestion (which breeds theft) would be inevitable because of the sheer number of different carriers that would be required each day to satisfy customer requests.

59. In areas where the rate of truck hijackings is high, suggests a police official, at each delivery point drivers should note the indicated mileage, leaving a record at the dock and in their log. If the truck is hijacked, the difference between the mileage recorded at the last delivery and the indicated mileage at the point of recovery-combined with other driver-supplied information-will assist police in pinpointing the drop or fence.

60. Rather than ship a fully assembled product, some companies ship the components and, in so doing, have reportedly reduced

cargo theft. For example, an encyclopedia publisher is said to ship odd-numbered volumes in one carton, even-numbered in another.

As indicated earlier, neither the above nor the many other possible procedures will achieve their full potential unless implemented as a well coordinated series of measures. That is, they should function as a system—not as unrelated, ad hoc controls.

Part Nine

TRANSPORTATION SECURITY SINCE SEPTEMBER 11, 2001

TRANSPORTATION SECURITY SINCE SEPTEMBER 11, 2001

When we were in the midst of sifting through what was a pile of rubble after the collapse of the World Trade Center towers, we didn't understand that within hours our national transportation infrastructures were also victims.

We did the right thing to embargo goods coming into the United States. We shut the borders down, we closed the airports and diverted air traffic, and we asked U.S. Customs to do the impossible by checking all cargo to eliminate weapons of mass destruction from entering the United States. An impossible task for U.S. Customs, considering that in the year 2000, 211,000 vessels, 127 million vehicles, 489 million people, over 6 million ship containers, and thousands of aircraft containing air shipments entered the United States. Sea, air, and border security was, and still remains in certain areas, inadequate. The fact that 9-11 happened is not a surprise in the open and free society that we enjoy in the United States. What is a surprise is that it took so long. An even bigger surprise is that more infrastructure attacks have not continued. We are walking on egg shells, waiting for the next event or series of events to take place.

We know that our transportation infrastructure has been in need of a security overhaul for thirty years or more. It has been inadequate in effectively reducing theft of goods in the transportation systems of all modes of intermodal transportation. We know we are vulnerable. We knew we were ill-prepared when 9-11 happened. Since then, our government has elected to throw unprecedented amounts of money at the problems—millions upon millions of dollars for seaports, airports, and border enforcement agencies. Even after months of priority security attention at our airports after 9-11, we read on the front pages of our newspapers that jet freight isn't "checked." The Transportation Security

Administration (TSA) of the Department of Transportation had not corrected security flaws even after having been given a report prepared by the Federal Aviation Administration. All of this was exposed by *USA Today* in May of 2002.

In 1999, the U.S. Department of Transportation reported some statistics of interest, which included the following:

- Aviation: 95 armed attacks and hijacks with 19 killed.
- Maritime: 65 attacks of which 12 were international terrorists with 90 killed.
- Piracy: 284 acts against merchant ships, most of which were in the Singapore Strait.
  (From the Rand Corporation and the Department of Homeland Security)
- Highways, bridges, and tunnels: The second most frequent transportation sector attacked with 632 killed and 281 injured by bombing and armed attacks. Of these, only seven were international attacks.

The implications of this are that everything is vulnerable. Our president has termed it a War—and it surely is. No longer does it only happen overseas; it happens in our homeland. We are victims and we are angry. What we must understand is that while we are at war, we haven't eliminated politics from its influence over how we attack the issues. Senator E.F. Hollings, (D-SC) wants to impose a tax on containers, calling it a user fee. Whatever you call it, the consumer will end up paying the cost to fight terrorism and improve security. Millions upon millions are being made available to study the problem. While we are reinventing the Cargo Security Wheel, the governmental process has stalled the passage of federal law requiring Smart ID cards, transportation workers background checks, hardware and technology improvements, intelligence sharing and numerous other considerations.

In April of 2002, the U.S. Customs Service introduced the Customs–Trade Partnership Against Terrorism (C-TPAT). This program will allow companies considered low-risk to streamline how they import goods. Low-risk companies would be those that already have an established process for monitoring the integrity of the cargo they import. Companies that would be accepted into the program would have an easier time getting goods into the country. U.S. Customs would be able to concentrate its limited resources on high-risk candidates. Companies must submit a request and document how they ensure security throughout their import process.* The rewards are many for participating in this program.

President Bush has developed the Office of Homeland Security. He seeks to galvanize the Immigration and Naturalization Service, Border Patrol, U.S. Customs Service, U.S. Coast Guard, and the Secret Service. Thus far, this has not been accomplished. The likelihood of it taking place remains in the hands of the Congress, where private political agendas and funding issues are many.

Legislative efforts in various states have enacted anti-terrorism laws and defined punishments. Among these are:

- New York: A.60002, S.70002, A.1452, and S.5646
- California: A.B. 1308
- Colorado: S.J.R.01S2-007
- Florida: Appointed 2 committees to draft laws
- Illinois: S.B.860-01 expands causes of disaster to include terrorism
- Kansas: Appointed committees
- New Jersey: A:9-64 Domestic Security Preparedness Act; A:9-67Domestic Security Preparedness
- Michigan: S.B.497
- Kentucky: H.B.1-01
- South Carolina: H.B.3223

  Note: Several other states have passed laws to support federal government responses, increase benefits for those victims directly affected by terrorism, and to express sympathy and support for the people of New York City and the Washington, D.C., area.

Internationally, several countries are engaged in reviews of appropriate laws dealing with domestic terrorism and acts of aggression by militant groups, but none more actively than the United Kingdom. The UK

* Coleman-Lochner, Lauren, *Bergen Record*, April 28, 2002.

passed the Terrorism Act, which provides for the following:

- Permanent UK-wide anti-terrorist legislation.
- A new definition of terrorism.
- New powers to seize suspected terrorist cash at borders.
- A new offense of inciting terrorist acts abroad from within the UK.

Related informational links:

Countering the Changing Threat of International Terrorism: Report of the National Commission on Terrorism
http://www.fas.org/irp/threat/commission.html

Justice Technology Information Network
http://www.justnet.org/techproj/

U.S. Department of Justice: Office of Justice Programs
http://www.ojp.usdoj.gov/terrorism/whats_new.htm

Terrorism Research Center
http://www.terrorism.com/index.shtml

Further information may be obtained from the National Conference of State Legislatures, Washington, D.C.: tel. no. 202-624-5400

Everyone concerned with global logistics security must actively support any and all changes that can improve integrity concerning the movements of goods and materials in the Global Supply Chain. Strict accountability, tougher physical security, accurate and confidential documentation transfers, effective government inspection and controls, and due diligence of vendors and suppliers of services all are important asset protection issues in the process of transportation.

APPENDICES

Appendix A

GLOSSARY

Key = (a) Air; (t) Trucking; (m) Maritime

ABI (m) U.S. Customs' automated broker interface, with which brokers file importers' entries electronically.

aboard (m) Refers to cargo being put, or laden, onto a means of conveyance.

access control facility (ACF) (m) Security system used by APC to protect unauthorized system entry and data access. Password protection and encryption, along with features built into communications procedures can authorize certain terminals to process only certain transactions.

accessorial charges (m) Charges that are applied to the base tariff rate or base contract rate, e.g. bunkers, currency, destination delivery charges.

accessorial services (m) A service in addition to the linehaul service, usually at added expense. Charges incurred to provide additional services at the time of pickup or delivery of the container. These additional services include driver assist in load or unload, lumper (additional labor), driver waiting time load or unload past free-time provision of the tariff, equipment rental (fork lift), blocking or bracing, additional stop-off charges, container charges, etc.

accessorial services (a) Services performed by an airline in addition to the normal transportation service. Common accessorial services include advancement of charges, pickup, delivery, C.O.D. service, signature service, and storage.

accessorial services (t) A service rendered by a carrier in addition to a transportation service, such as stopping in transit to complete loading or to partially unload, heating, storage, etc.

account number (a) An identifying number issued by carrier accounting offices to identify a shipper and/or consignee. The number ensures accurate invoicing procedures and customer traffic activity.

account party (m) The party instructing the bank to open a letter of credit on whose behalf the bank agrees to make payment. In most cases, the account party is an importer or buyer, but may also be a construction contractor or a supplier bidding on a contract.

acquiescence (m) When a bill of lading is accepted or signed by a shipper or shipper's agent without protest, the shipper is said to acquiesce to the terms, giving a silent form of consent.

acquittance (m) A written receipt in full, in discharge from all claims.

active seal A seal that is electrically powered and often reusable.

act of God (m) An act beyond human control, such as lighting, flood, or earthquake.

act of God (t) An irresistible superhuman cause, such as no reasonable human foresight, prudence, diligence and care can anticipate and/or prevent.

act of man (m) In water transportation, the deliberate sacrifice of cargo to make the vessel safe for the remaining cargo. Those sharing in the spared cargo proportionately cover the loss.

Act to Regulate Commerce (t) An Act of Congress regulating the practices, rates, and rules of transportation lines engaged in handling interstate traffic (now known as the Interstate Commerce Act).

adhesive tape seal Sticky label that is damaged if removed from what it adheres to. Often used as tags. Also known as adhesive label seal or pressure-sensitive adhesive seal.

administrative law judge (m) A representative of the Interstate Commerce Commission or other agency vested with power to administer oaths, examine witnesses, take testimony, and otherwise conduct hearings of cases submitted to, or initiated by, that agency. Formerly called hearing examiner.

adoption notice (t) A notice required to be filed with the Interstate Commerce Commission by a carrier or person taking over operating control of another carrier.

ad valorem (t) According to value. Refers to taxes on property.

advance arrangements (a) The shipment of certain classes of commodities in air freight—examples: gold, precious gems, furs, live animals, human remains, and oversized shipments—require arrangements in advance with carriers. For detailed information consult any airline.

advanced charge (t) Any charge other than the applicable freight charges advanced by one transportation line to another, or to the shipper, to be collected from the consignee.

advancement of charges (a) A service under which airlines, in some instances, pay incidental charges arising before or after an air haul. Examples would include cartage and warehousing costs. These charges can be in advance for the convenience of either the shipper or their receiver.

advice of shipment (m) A notice sent to a local or foreign buyer advising that shipment has gone forward and containing details of packing, routing, etc. A copy of the invoice is often enclosed and sometimes, if desired, a copy of the bill of lading.

advised credit (m) A letter of credit that has been advised by another bank without engagement on the part of that bank. An advising bank has the obligation to take reasonable care to check the apparent authenticity of the credit, which it advises.

advising bank (m) A bank, operating in the exporter's country, that handles letters of credit for a foreign bank by notifying the export firm that the credit has been opened in its. favor. The advising bank fully informs the exporter of conditions of the letter of credit without necessarily bearing responsibility for payment.

advisory capacity (m) A term indicating that a shipper's agent or representative is not empowered to make definite decisions or adjustments without prior approvals of the group or individual represented.

affiant (t) One who makes an affidavit.

affidavit (m) A written statement generally sworn to before a Notary Public.

affidavit (t) A written statement sworn to before a Notary Public.

agencies (m) The right given by an individual, body, or firm to another person or company to act on its behalf. An agency may be classified as: a specific or special agency where duties are restricted to a specified transaction; a general agency where actions are undertaken on behalf of principals, with an amount of responsibility; and a universal agency which involves full power of attorney provided all actions are within the law. Agency appointments are normally given for a stated period of time or indefinitely by means of a written contract or letter of intent. There is wide scope for the activity of an agency, such as representing the ship owner, contracting agreements for repairs and dry-docking, acting in the absence of the owners, or importing materials.

agency (t) Used interchangeably with *call station* but generally an agent picks up and delivers freight for an intercity carrier while a call station handles orders from shippers but performs no freight handling functions.

agency tariff (t) A tariff issued by a publishing agent for one or more transportation lines.

agent (t) A person authorized to transact business for and in the name of another.

agent (m) A person or company acting as an agency, authorized to act on behalf of its principal. The agent is very important in that he will be the owner's representative who has to act promptly and effectively to deal with problems. The agent is duty bound to follow the owner's instructions and may indicate on whose authority he is acting. There are many levels or representation characterized as agent, including managing agent or managing owner, which may actually be the owner of the ship. Typically the agent handles all formalities for a ship's entry into port, during customs clearance, and up to time of sailing. The agent arranges with the authorities for the allocation of berthing space, advises cargo owners of the availability of the ship or cargo (depending on whether the cargo is for export or import), organizes loading and unloading, and provides for provisioning through a ship chandler.

aggregated shipments (t) Several shipments from different shippers to one consignee that are consolidated and treated as a single consignment.

agreed valuation (t) Freight value mutually agreed upon by shipper and carrier as a basis for transportation charges. It also may represent an agreed maximum amount that can be recovered in case of loss or damage.

agreed valuation (m) The value of a shipment agreed upon in order to secure a specific freight rate.

agreed weight (t) A weight set and agreed upon by carrier and shipper as a basis for charges.

agreed weight (m) The weight prescribed by agreement between carrier and shipper for goods shipped in certain packages or in a certain number.

airbill (a) *See air waybill.*

air cargo (a) Any property (freight, mail, express) carried or to be carried in an aircraft. Does not include passenger baggage.

air cargo handling area (a) Those portions of an air terminal designated and made available by the leasing facility for the reception, storage, and distribution of air cargo or for the loading and unloading of air cargo from all cargo aircraft. Also known as a ramp area.

air cargo employee (a) A person employed by an air cargo operator who performs duties in the air cargo handling area.

Air Cargo Guide (a) A basic reference publication for shipping freight by air. It contains current domestic and international cargo flight schedules, including pure cargo, wide body, and combination passenger-cargo flights. Each monthly issue also contains information on air carriers' special services, labeling, airline and aircraft decodings, air carrier and freight forwarders directory, cargo charter information, United States and Canadian city directory small package services, interline air freight agreements, aircraft loading charts and more.

Air Cargo, Inc. (ACI) (a) A ground service corporation established and jointly owned by the United States scheduled airlines. In addition to its airline owners, ACI also serves over 30 air freight forwarders and international air carriers. One of ACI's major functions is to facilitate the surface movement of air freight by negotiating and supervising the performance of nationwide series of contracts under which trucking companies provide both local pickup and delivery service at airport cities and over-the-road truck service to move air freight to and from points not directly served by the airlines. ACI publishes a directory of these trucking services, listing over 19,000 points served in the United States and the applicable pickup and delivery rates. Other services include claims inspection, terminal handling, telemarketing service, and group purchasing (equipment, supplies, insurance). ACI also makes available, in many cities, low cost, disposable containers for shippers' use.

air cargo operator (a) Any person who has a lease or other agreement with the facility authorizing or consenting to the use of any portion of the air cargo handling area.

aircraft pallet (a) The use of a platform or pallet in air freight (usually from ¾-inch to 2 inches thick) on which unitized shipment rests and where goods are assembled and secured before being loaded as a unit onto the aircraft. Most carriers offer container discounts for palletized loads. Palletization results in more efficient use of space aboard freighter aircraft and better cargo handling, particularly when used as part of mechanized systems employing such other advances as pallet loaders and pallet transporters.

aircraft pallet loader (a) A device that uses one or more vertical lift platforms for the mechanical loading or unloading of palletized freight at planeside.

aircraft pallet transporter (a) A vehicle used for the movement of loaded pallets between the aircraft and the freight terminal or truck dock. Sometimes the functions of both the pallet loader and pallet transporter are combined into a single vehicle. *See pallet.*

air express (a) A term often used to describe expedited handling of airfreight service. See priority air freight.

air freight (a) A service provided for the transport of goods in any volume. Air freight is, by far, the predominate form of air cargo accounting for over 80 percent of all air cargo movement. Each shipper's decision to move goods by air turns on one or more of these basic factors: time, inventory control, superior condition of goods upon arrival, and lower shipping costs. Increasingly, the decision to ship by air is a market oriented decision and hence, a consumer oriented decision.

air freight forwarded (a) Serves a dual role. The air freight forwarders are, to the shipper, an indirect carrier because they receive freight from shippers under their own tariff usually consolidating it into larger units which is tendered to the airlines. To the airlines, the air freight forwarder is a shipper. An air freight forwarded is ordinarily classed as an indirect air carrier; however, many air freight forwarders operate their own aircraft.

airmail (a) The term "airmail," as a class of mail, is used only in international postal service. Within the United States, the U.S. Postal Service moves all first class mail, priority mail, and express mail by air where this will expedite delivery.

air parcel post (a) A term commonly used for priority mail which consists of first class mail weighing more than 13 ounces. Priority mail is another economical and expedited service for the shipping of parcels by air.

air waybill (airbill) (a) Shipping document used by the airlines for airfreight. It is a contract for carriage that includes carrier conditions of carriage including such items as limits of liability and claims procedures. The air waybill also contains shipping instructions to airline, a description of the commodity, and applicable transportation charges. Air waybills can be used by many truckers as through documents for coordinated air/truck service. Air waybills are not negotiable. The airline industry has adopted a standard formatted air waybill that accommodates both domestic and international traffic. The standard document was designed to enhance the application of modern computerized systems to air freight processing for both the carrier and the shipper.

air waybill (m) The air waybill (also called air consignment note) is the forwarding agreement or carrying agreement between shipper and air carrier and is obtained from the airline used to ship the goods in question. Air waybills are issued only in non-negotiable form.

air-cargo aircraft (a) An aircraft for the carriage of cargo only, rather than the combination of passengers and cargo. The all-cargo aircraft will carry traffic in bulk or container in the main deck as well as in the lower deck of the aircraft. It may include a scheduled and nonscheduled service.

allocate (t) To assign or allot.

allowance (t) 1. A sum granted as a reimbursement or repayment. 2. A deduction from the gross weight or value of goods.

all-risk clause (m) An insurance provision that all loss or damage to goods is insured except that of inherent (self-caused) vice.

alternative rate (t) The rate employed where tariffs provide for the use of more than one published rate on some particular traffic.

alternative tariff (t) An article not found in a freight classification but having similar characteristics to one in the classification.

analogous articles (t) An article not found in a freight classification but having similar characteristics to one in the classification.

ancillary equipment (a) Equipment used to build up a palletized load or to convey a unit load device outside an aircraft. *See aircraft pallet.*

antipilferage seal (a) seal used primarily to detect theft and possibly discourage it.

any-quantity rate (t) A rate applicable to an article regardless of how much is to be transported.

apparent good order (m) When freight appears to be free of damage so far as a general survey can determine.

appraisement (m) Determination of the dutiable value of imported merchandise by a Customs Service official who follows procedures outlined in their country's tariff, such as the U.S. Tariff Act of 1930, as amended.

appraiser's stores (m) The warehouse or public stores to which at least ten percent of imported goods are taken to be inspected, analyzed, weighed, etc. by examiners or appraisers.

arbitrary (t) 1. A fixed amount which a transportation line agrees to accept in dividing joint rates. 2. A fixed amount added to or deducted from a rate from one station to make a rate for another station. 3. A fixed amount added to or deducted from a rate to one station to make a rate to another station.

arrest (m) An official order enforced as a matter of law against a ship for actions against the laws of the government ordering the arrest, or as a means of enforcing a maritime lien.

arrival notice (m) A notification by carrier of ship's arrival to the consignee, the "Notify Party," and when applicable the "also Notify Party." These parties in interest are listed in blocks 3, 4, and 10, respectively of the Bill of Lading.

arrival notice (t) A notice, furnished to consignee, of the arrival of freight.

articles of extraordinary value (AEV) (a) Commodities identified as high value items.

assembly service (a) A service under which an airline assembles shipments from many shippers and transports them as one shipment to one receiver.

assignment (m) A term commonly used in connection with a bill of lading. It involves the transfer of rights, title and interest in order to assign goods by endorsing the bill of lading.

assignment of proceeds (m) A request by the beneficiary to pay all or part of the funds due from a third party.

astray freight (m) Lost freight that has been found.

astray freight (t) Freight bearing marks indicating origin and destination, but separated from the waybill. *See over freight.*

astray freight waybill (m) A waybill used to move astray freight to its proper location without charges.

attendant accompanying shipments (a) Sometimes attendants accompany cargo shipments as when grooms or veterinarians accompany race horses or other live animals. This service requires advance arrangements with an airline.

audit of freight bills (t) The process of verifying the correctness of the transportation charges shown on the carrier's freight bill.

authority to pay (m) (Often mistaken for letter of credit) An authority to pay is an advise stemming from a buyer, addressed through the buyer's bank to the seller, by way of the correspondent bank to pay the seller's drafts for a stipulated amount. The authority to pay may be cancelled or modified before the drafts are presented. Once the drafts drawn on the correspondent bank are paid by it, the seller is no longer liable as drawer. Usually not offered by U.S. banks.

authority to purchase (m) This document, while similar to the authority to pay, differs in that under an Authority to Purchase the drafts are drawn directly on the buyer. They are purchased by the correspondent bank with or without recourse against the drawer. Usually not confirmed by the U.S. bank. Authority to purchase is used primarily in the Far East.

automated commercial systems (ACS) (m) U.S. Customs' master computer system, "Automated Commercial Systems."

automatic POD (a) formation automatically sent to payor containing name of person who signed for the package with date and time of delivery.

automatic proof of delivery (a) Same as automatic POD.

awaiting orders (m) A ship is awaiting orders when it has not been instructed where it is to discharge its cargo, or when its next employment is undermined. A ship legitimately declaring that it is awaiting orders may be underway (but usually near a discharge port), at anchor in a holding area, or adrift on the high seas outside normal shipping lanes.

back haul (t) 1. Traffic moving in direction of light flow when a carrier's traffic on a route is heavier in one direction than the other. 2. To haul a shipment back over a part of a route it has traveled.

back-to-bank letters of credit (m) A letter of credit supported by ("backed by") a separate letter of credit having nearly identical documentary requirements, but with four features normally differing: the name of the beneficiary; the account party, the amount; and the shipping date.

ballast, in-ballast (m) A ship is in ballast when it is empty of cargo, but has an amount of (usually) sea water in ballast tanks. This serves to keep the propeller and rudder effectively submerged, and to keep the bows in the water to improve handling and prevent slamming. A ship in ballast may or may not be evenly trimmed fore and aft, but will never normally be ballasted so deeply as to approximate the loaded condition; load line at or near the water line. A merchant ship on a long voyage will not normally ballast down to replace fuel consumed. Tankers and bulk carriers will normally sail either in an obviously light (ballasted) condition or down to the load line. However, chemical tankers and those carrying refined products will often sail partially loaded. Tankers out of Red Sea ports may sail less than full in order to meet draft.

balloon freight (t) Light, bulky cargo.

bank draft (m) A check drawn by a bank on another bank. This is customarily used when it is necessary for the customer to provide funds which are payable at a bank in some distant location.

banker's acceptance (m) A form of credit created when a bank acknowledges in writing on a time draft its obligation to pay face amount to the holder at a specific time in the future.

banker's bill (m) A bill of exchange drawn by an exporter on the importer's bank.

bareboat charter (m) Also known as demise charter. A charter for a long period, where the charterer assumes responsibility for the normal functions of a ship owner; i.e., manning and appointment of officers, payment of running expenses, dry docking, painting, and repair. In some cases, the registry and name of the ship will also be changed. In a bareboat charter, the powers of ownership have been assumed by the charterer, regardless of the registered or beneficial owner of record. In many instances the bareboat charterer becomes the outright owner at the end of the charter period. In others, the owner of the ship through a foreign "paper" company will bareboat charter the ship back to himself, gaining the benefits of control over a ship without having to comply with his own national rules regarding manning and operations. This form of bareboat charter is most common with German owners using Cypriot or Filipino front companies.

barrier seal A security device that functions both as a lock and a seal. Also called a hybrid seal.

basing point (t) A point upon which through rates to another destination are computed: For example, a rate from Louisville, Kentucky to a point near Jacksonville, Florida is computed as follows: The rate from Louisville to Jacksonville, plus the local rate from Jacksonville to the nearby point. Jacksonville, in this case, is a basing point.

basing rate (t) *See proportional rate.*

beneficiary (m) The party who receives payment upon the conditions stipulated in a letter of credit. This party is usually a seller or an exporter. The person in whose favor a letter of credit is issued or a draft is drawn.

billed weight (t) The weight shown in a freight bill.

bill of lading (a) A document by which a carrier receipts for goods and contracts to move them. In air freight, the air waybill is a form of bill of lading and is the contract of carriage.

bill of lading (t) The written transportation contract between shipper and carrier (or their agents). It identifies the freight, who is to receive it and the place of delivery and gives the terms of the agreement. All goods going to a receiver at one destination in a single shipment or on one truck must be on a single bill of lading. The straight bill of lading provides that freight be delivered to the receiver shown on the bill. It is non-negotiable. Surrender of this type of bill of lading is not required except when it is necessary for the receiver to identify himself. This might be necessary for certain bonded goods such as liquor. It is always printed on white paper. The order bill of lading is negotiable. Its purpose is to enable a shipper to collect for the shipment before it reaches its destination. The shipper sends an original bill of lading and a draft for the charges through a bank. The receiver pays the carrier's agent the amount of the draft and then can receive the goods. With this method, the shipper customarily consigns the shipment to himself. The person or company to be notified at destination is specified. The shipment may be released to the receiver only upon the order of the shipper. The order bill of lading must be surrendered with delivery of the freight. Always printed on yellow paper, either straight or order bills of lading may be designated as "through." A through bill of lading covers shipments by more than one transportation company at a fixed rate for the entire service. More than one type of transportation company (as truck and rail) may be used.

bill of lading (B/L) (m) A document that establishes the terms of a contract between a shipper and a transportation company. It serves as a document of title, a contract of carriage, and a receipt for goods.

amended B/L: B/L requiring updates that do not change financial status; this is slightly different from corrected B/L.

B/L terms and conditions: the fine print on back of B/L; defines what a transportation company can and cannot do, including liabilities and contractual agreements.

B/L's status: represents whether the bill of lading has been input, rated, reconciled, printed, or released to the customer.

B/L's type: refers to the type of B/L being issued. Some examples are: a memo (ME), original (OBL), non-negotiable, corrected (CBL) or amended (AM) B/L.

canceled B/L: B/L status; used to cancel a processed B/L; usually per shipper's request; different from voided B/L.

clean B/L: a clean B/L is one which bears no superimposed clause or notation which expressly declares a defective condition of the goods and/or the packaging.

combined B/L: B/L that covers cargo moving over various transports.

consolidated B/L: B/L combined or consolidated from two or more B/Ls.

corrected B/L: B/L requiring any update which results in money or other financially related charges.

domestic B/L: non-negotiable B/L primarily containing routing details; usually used by truckers and freight forwarders.

duplicate B/L: Another original bill of lading set if first set is lost. Also known as reissued B/L.

express B/L: Non-negotiable B/L where there are no hard copies of originals printed.

freight B/L: A contract of carriage between a shipper and forwarder who is usually a NVOCC; a non-negotiable document.

government B/L (GBL) A document issued by the U.S. military.

hitchment B/L: B/L covering parts of a shipment which are loaded at more than one location. Rule: one ship, one shipper, one consignee, one destination, one B/L made of two parts-hitchment plus hitchment memo. Hitchment usually covers majority portion of divided shipment and carries the entire revenues.

house B/L: A B/L issued by a freight forwarder or consolidation covering a single shipment containing the names, addresses and specific description of the goods shipped.

memo hitchment B/L: Memo covering balance of shipment. Normally no freight money amount is assigned to the memo.

intermodal B/L: B/L covering cargo moving via multimodal means. Also known as combined transport B/L, or multimodal B/L.

"local" B/L: Familiar term for a B/L accommodating cargo covered by a local tariff (i.e., non-land-bridge).

long form B/L: B/L form with terms and conditions written on the back.

memo B/L: B/L to cover cargo when a commercial B/L is not needed.

military B/L: B/L issued by that U.S. military; also known as GBL, or Form DD1252.

multimodal B/L: B/L covering cargo moving intermodally also known as combined transport B/L, intermodal B/L.

B/L numbers: U.S. Customs' standardized B/L format to facilitate electronic communications.

NVOCC B/L: B/L issuer is a non-vessel-operating common carrier, e.g., a consolidator or forwarder.

negotiable B/L: Consignment/banking term. B/L names legal and by endorsement the shipper can transfer the title of the goods to the bank representing the buyer or directly to the buyer of the goods.

non-negotiable B/L: *See straight consignment.* Term is also used sometimes to mean file copies of B/L.

ocean B/L: As opposed to domestic B/L. Archaic; not used much.

onboard B/L: B/L validated at the time of "onboard" (to a transport).

onboard air B/L: B/L validated at time cargo is loaded to aircraft.

onboard boxcar B/L: B/L validated at time cargo is loaded to boxcar.

onboard container B/L: B/L validated at time cargo is loaded to container.

onboard rail B/L: B/L validated at time cargo is loaded to train.

onboard truck B/L: B/L validated at time cargo is loaded to truck.

onboard vessel B/L: B/L validated at time cargo is loaded to ship or vessel.

optional discharge B/L: B/L covering cargo with more than one discharge point option possibility.

order B/L: *See to order B/L or negotiable B/L.*

original B/L (OBL) The part of the B/L set that has value, especially when negotiable; rest of set are only informational copies.

received for shipment B/L: B/L validated at time cargo is received by international carrier to commence movement.

reconciled B/L: B/L set which has completed a prescribed number of edits between the shipper's instructions and the actual shipment received. This produces a very accurate B/L.

reissued B/L: *See duplicate B/L.*

short-term B/L: Opposite of long form B/L, a B/L without the terms and conditions written on the back.

short shipped B/L: Familiar name given to a B/L covering only a portion of an entire shipment; balance to be covered by a memo B/L.

split B/L: One of two or more B/Ls which have been split from a single B/L.

Stale B/L: A late B/L; in banking, a B/L which has not met the time deadline of the L/C and is therefore considered useless.

straight (consignment) B/L: Consignment issue. *See non-negotiable B/L.*

"to order" B/L: Consignment issue. *See negotiable B/L.*

unique B/L identifier: U.S. Customs: standardization: four-alpha code unique to each carrier placed in front of nine-digit B/L number.

unique B/L number: *See nine-digit B/L number.*

voided B/L: Related to consolidated B/L; those B/Ls absorbed in the combining process. Different from cancelled B/L.

Bills of Lading Act (t) A 1917 Act of Congress pertaining to the preparation and negotiability of bills of lading.

bill of sale (t) A contract for the sale of goods.

bill of sale (m) confirms the transfer of ownership of certain goods to another person in return for money paid or loaned.

billed weight (m) The weight shown in a waybill and freight bill.

billing information (m) This term is used to describe the gathering of information from partner carriers. The elements of billing information are as follows: pieces, weight, bill of lading, number, seal, Hazmat information, pick up and delivery times.

billing third party (a) The invoicing of transportation charges to other than shipper or consignee.

blanket bond (t) A bond covering a group of persons, articles or properties.

blanket rate (t) 1. A rate applicable from and/or to a group of points. 2. A special rate applicable on several different articles in a single shipment.

blanket bond (m) A bond covering a group of persons, articles or properties.

blanket tariff supplement (t) A single publication containing additions to or changes in two or more tariffs.

blocked accounts (m) Deposits maintained in a country that does not allow conversion into another currency or removal of the deposits from the country.

blocking (t) Supports used to prevent shipments from shifting during transportation.

blue label (t) Atomic material shipment.

bob-tail (t) Tractor operating without a trailer. Also a straight truck.

bolt seal A barrier seal consisting of a strong bolt with one end larger than the hasp and the other end designed to snap irreversibly into a cylindrical head or housing. Especially popular for use on trucks or transportainers.

bona fide (t) In good faith; authentic, genuine.

bond of indemnity (t) An agreement made with a transportation line relieving it from liability for which it would otherwise be liable.

bonded freight (m) Freight moving under a bond to U.S. Customs or to the Internal Revenue Service, and to be delivered only under stated conditions.

bonded indemnity (m) A certificate filed with a carrier, relieving it from liability to which it would otherwise be subject.

bonded terminal (a) An airline terminal approved by the U.S. Treasury Department for storage of goods until Customs duties are paid or the goods are otherwise released.

bonded warehouse (a) A warehouse authorized by Customs authorities for storage of goods on which payment of duties is deferred until the goods are removed.

bonded warehouse (m) A warehouse authorized by Customs authorities for storage of goods on which payment of duties is deferred until the goods are removed.

bonded warehouse (t) A warehouse owned by persons approved by the Treasury Department. Under bond or guarantee for the strict observance of the revenue laws, they are used for storing goods until duties are paid or goods are otherwise properly released.

booking (m) An arrangement with a steamship company for the acceptance and carriage of freight. Generally, the first point of contact between the shipper and carrier. Acts as a reservation to hold space on a vessel.

boot topping (m) Technically, the area of a ship's hull between the light and full load waterline. Typically, the paint in the area of the ship's hull, which is alternately immersed and exposed as state of loading changes. The boot topping need not stop at, or near, the load waterline, and the amount of exposed boot topping, alone, cannot provide an accurate measure of the state of ship's loading.

bracing (t) *See blocking.*

break bulk (a) Disassembling of consolidated shipment for delivery or re-consignment of a shipment..

break bulk (m) To unload and distribute a portion of all of the contents of a rail car, container, or trailer. "Loose" non-containerized cargo.

break bulk (t) To separate a composite load into individual shipments and route to different destinations.

break bulk point (t) A place where composite load is separated into individual shipments and routed to different destinations.

brief (t) A written abstract of testimony and pleadings in a case and commentaries thereon.

broken up (m) Said of a ship when it has been reduced to scrap metal. Ships may be removed from service and dismantled, but remain in existence as floating storage or in service as barges. Ships sold to be broken up may continue in service for indefinite periods of time. It is very common for a ship to carry a final cargo, often of scrap metal, on its delivery to the breakers.

bunkers (m) Fuel consumed by the engines of a ship, or the, compartments in which such fuel is carried.

broker (customs) (a) A person or firm licensed by the U.S. Treasury Department to transact business with Customs on behalf of importers.

brokerage license (t) Authority granted by Interstate Commerce Commission to persons to engage in the business of arranging for motor vehicle transportation of persons or property in interstate commerce.

bulk freight (t) Freight not in packages or containers.

bulk freight (m) Not in packages or containers; shipped loose in the hold of a ship. Grain, coal and sulfur are usually bulk freight.

bulk-freight container (m) Refers to a container with a discharge hatch in the front wall; allows bulk commodities to be grasped by loading hatches.

burden of proof (m) The duty of proving a particular position in a legal proceeding. Failure to perform this duty calls for judgment against the party on whom the duty is imposed.

burden of proof (t) The obligation to prove disputed facts at issue in a legal proceeding. For instance, the Interstate Commerce Act provides that the burden of proof is upon the carriers to show that change in rates, rules, etc., are reasonable.

buyer credit (m) A credit arrangement established for financing of imported products with the buyer carrying the credit obligation. Recourse is to the buyer and the strength of the credit rests on the credit worthiness of the buyer. Export guarantee and insurance programs from the country are often available to facilitate such transactions.

cable seal A barrier seal made from aircraft cable with each end crimped or irreversibly clamped into a head, housing, or "locking" body.

cable transfer (m) A type of remittance using electronic transmission to move funds from one bank to a named party at another bank.

cabotage (m) Water transportation term applicable to shipments between ports of a nation; commonly refers to coast-wide navigation or trade. Many nations, like the United States, have cabotage laws that require domestic vessels to provide domestic interport service.

CAD (m) A term used to mean "clear at discharge." The shipment must clear Customs at the discharge post. Must be full container, approved by discharge/destination logistics if CFS shipment to be reworked at discharge port.

camber (m) Curvature of a deck such that it is highest fore and aft along the centerline and curves downward at the sides. Provides strength and facilitates rue-off of waves.

camet (m) A Customs document permitting the holder to temporarily carry or send merchandise into certain foreign countries (for display, demonstration or similar purposes) without paying duties or posting bonds.

captain's protest (m) A document prepared by the captain of a vessel on arriving at port; shows conditions encountered during voyage, generally for the purpose of relieving ship owner of any loss to cargo and shifting responsibility for reimbursement to the insurance company.

captive register (m) A register of ships maintained by a territory, possession, or colony primarily or exclusively for the use of ships owned in the parent country. Also referred to as an offshore register, and the "offshore" equivalent of an internal register. Ships on a captive register will fly the same flag as the parent country, or a local variant of it, but will be subject to the maritime laws and taxation rules of the offshore territory. Although the nature of a captive register makes it especially desirable for ships owned in the parent country, like the internal register the ships may also be owned abroad. The captive register then acts as a flag of convenience register, except that it is not the register of an independent state.

cargo (m) Freight loaded into a ship

cargo (t) The freight carrier by a vehicle.

cargo aircraft (a) *See all-cargo aircraft.*

cargo agent (a) An agent appointed by an airline to solicit and process international air freight for shipments. Cargo agents are paid commissions by the airline.

cargo manifest (m) A manifest that lists only cargo, not charges.

cargo NOS (m) Cargo not otherwise specified. Usually dangerous and hazardous (D&H) cargo which does not have a specific proper shipping name. Usually the first rate entry in a tariff that can apply to commodities not covered under a specific item or sub-item in the applicable tariff.

cargo terminal (a), (t), (m) An assigned area in which cargo (containerized or non-containerized) is prepared for loading into a vessel, train, truck, or airplane or is stacked immediately after discharge from the vessel, train, truck, or airplane; any area assigned for loading/unloading, temporary storage of vehicles, or the interchange of cargo during transit.

cargo tonnage (m) The weight ton in the United States is the short ton, 2,000 pounds, or the long ton, 2,240 pounds. In British countries, it is the English long ton, or gross ton, of 2,240 pounds. In France and other countries having the metric system, the weight ton is 2,204.62 pounds. The measurement ton is usually 40 cubic feet, but in some instances a larger number of cubic feet is taken as a weight ton. Most ocean freight is taken at weight or measurement (W/M), ship's option.

Carmack Amendment (t) An amendment to the Interstate Commerce Act which specifies carrier legal liability in connection with the bill of lading. (Section 20 of the Act).

carnet (a) A customs document permitting the holder to carry or send merchandise temporarily into certain foreign countries (for display, demonstration, or similar purposes) without paying duties or posting bonds.

carrier (t) An individual, partnership of corporation engaged in the business of transporting goods or persons.

carrier's lien (m) Right of carrier to retain property as security for charges.

carrier's lien (t) Carrier's claim on property it has transported as security for charges.

cartage (m) Usually refers to intra-city hauling on drays or trucks.

cartage (local) (t) Hauling between locations in the same town or city or contiguous municipalities.

cartage agent (a) Ground service operator who provides pickup and delivery in areas not served directly by air carrier.

cartage ticket (m) Customs form permitting in-bond cargo to be moved from one location to another under Customs control, within the same Customs district. Usually in motor carrier's possession while draying cargo.

cash against documents (CAD) (m) A method of payment for goods in which documents transferring title are given the buyer upon payment of cash to an intermediary acting for the seller, usually a commission house.

cash in advance (CIA) (m) A method of payment for goods in which the buyer pays the seller in advance for the shipment of goods.

Usually employed when the goods, such as specialized machinery, are built to order.

cash with order (CWO) (m) A method of payment for goods in which cash is paid at the time of order and the transaction becomes binding on both buyer and seller.

cellular vessel (m) Ship specially designed and arranged for the carriage of containers. Holds or cells are arranged so that the containers are lowered and stowed in a vertical line and restrained at all four corners by vertical posts. Normally stowed containers sit up to seven high below decks and three to four high above decks.

certificate of inspection (m) A document certifying that merchandise (such as perishable goods) was in good condition immediately prior to its shipment. Also, the document issued by the U.S. Customs Guard certifying an American Flag vessel's compliance with applicable laws and regulations.

certificate of manufacture (m) A statement (often notarized) in which a producer of goods certifies that manufacture has been completed and that the goods are now at the disposal of the buyer.

certificate of origin (a) A document required by U.S. Customs and certain other countries for tariff purposes certifying as to the country of origin of specified goods.

certificate of origin (m) A certified document showing the origin of goods; used in international commerce.

certificate of public convenience and necessity (t) Authority or certificate granted by the Interstate Commerce Commission or state regulatory agencies and required of certain for-hire carriers.

certificate of weight (t) An authoritative statement of the weight of a shipment prepared by a weighmaster.

CGS-container freight station (m) A warehouse facility, either on-dock or off-dock, where less-than-container load (LCL) shipments are consolidated into full container load shipments.

chandler (m) A merchant who supplies ships with stores and provisions.

chargeable weight (a) The weight of the shipment used in determining air freight charges. The chargeable weight may be the dimensional weight or on container shipments the gross weight of the shipment less the tare weight of the container.

charges advanced (a) *See advancement of charges.*

charges collect (a) Transportation charges may include pickup and/or delivery and are entered on the air waybill to be collected from the consignee. Equivalent terms are "freight collect" or "charges forward."

charges for packing shipments (a) Some carriers provide material and services for packing and unpacking special commodities that require special packaging and handling. A charge is made for this service.

charges (payment of transportation charges) (t) Ordinarily, unless specified arrangements have been made, no carrier is permitted to deliver or relinquish possession at destination of any freight transported by it until all tariff rates and charges thereon have been paid. Carriers upon taking precaution deemed to be sufficient to assure payment of tariff charges may relinquish possession of freight in advance of payment of tariff charges and may extend credit in the amount of such charges for a period of seven days.

charter (m) The hiring of a vessel or part of its capacity. Chartering activity is one of the least reported of legitimate maritime operations; most deals being made on a private and confidential basis. Only approximately 50 percent of charters are ever reported in commercial shipping sources. Charters of less than full ship (space charter, slot charter) are almost never reported.

charter party (m) The basic document between vessel owner and charterer setting forth the type charter, duration, and rates of payment, tonnage, capacity and condition of the vessel course it is to pursue, loading and discharge ports, redelivery port, responsibility for bunkering, insurance, manning, and operations.

chartered ship (m) A vessel under lease by its owner to others.

chassis (m) A trailer constructed to accommodate containers which are moved over the road.

check digit number (a) A single digit of the air waybill number used to insure that the air waybill number is correctly entered into a computer system.

CHIPS (m) An acronym for the clearinghouse interbank payment system, a computer system operated by New York banks to settle international payments. Using CHIPS, checks are cleared, other instruments are exchanged and net balances are settled among banks.

citation of decision (t) A reference made to the location of a decision or opinion rendered by a court or a commission. A citation of an Interstate Commerce Commission decision could be to a specific page and volume of the Interstate Commerce Commission's Report.

circuitous route (t) An indirect route.

city terminal service (a) A service provided by some airlines to accept shipments at the terminals of their cartage agents or other designated in-town terminals or to deliver shipments to these terminals at lower rates than those charged for the door-to-door pickup and delivery service.

claim (m) A demand made upon a transportation line for payment on account of a loss sustained through its negligence.

claim (t) 1. A demand made upon a transportation line for payment on account of loss or damage alleged to have occurred while shipment was in possession or carrier. 2. A demand upon a transportation company for refund of an overcharge. *See overcharge.*

claim agent (t) An employee who adjusts or settles claims made against his or her company.

claimant (t) Person or company filing a claim.

claim tracer (m) A request for advice concerning the status of a claim.

claim tracer (t) A request for information concerning the status of a claim.

Class I motor carriers (t) Common or contract motor carriers of property that have average gross operating revenues (including interstate and intrastate) of $1,000,000 or more annually from motor carrier operations. For passenger carriers, the monetary limit is $200,000.

Class I motor carriers (m) Motor carriers with average annual operating revenue of $1,000,000 or over for the three immediately preceding years.

Class I railroads (m) Those railroads with an annual operating revenue of $5,000,000 or more.

Class II motor carriers (m) Motor carriers with average annual operating revenue between $3,000,000 and $1,000,000 for the three immediately preceding years.

Class II motor carriers (t) Common or contract motor carriers of property that has average gross operating revenues (including interstate and intrastate) of $200,000 or more, but under $1,000,000 annually from motor carrier operations. For passenger carriers, it is more than $50,000 but less than $200,000.

Class II railroads (m) Those railroads with annual operating revenue of less than $5,000,000.

Class III motor carriers (m) Those motor carriers with average annual operating revenue of less than $300,000 for the three immediately preceding years.

Class III motor carriers (t) Common or contract motor carriers of property that have average gross operating revenues (including interstate and intrastate) of less than $200,000 annually, from motor carrier operations. For passenger carriers, it is less than $50,000.

class rate (t) A transportation charge set for a group of commodities. Unless an article is given a special freight rate, it is grouped with others of a similar nature into a class. A transportation charge is set for the class. This charge or rate applies to each type of article in the class. *See classification rating.*

classification (m) A publication, such as the *Uniform Freight Classification* (railroad) or the *National Motor Freight Classification* (motor carrier), that assigns ratings to various articles and provides bill of lading descriptions and rules.

classification (t) A publication containing freight ratings. *See classification rating.*

classification rating (t) The class to which an article is assigned for the purpose of applying transportation charges. The rating expresses the mathematical relationship of the freight rate for the group to a base rate, called the first class rate. The class ratings are expressed as percentages of first class if they are less than the first class rate. If higher than the first class rate, they are called multiples. For example, a class 70 rate would be 70 percent of first class. A rate of 150 would be 1.5 times the first class rate.

class tariff (t) A published schedule of rates that contains only class rates.

class and commodity tariff (t) A tariff containing both class and commodity rates.

clean bill of lading (t) A bill of lading signed by the carrier for receipt of merchandise in good condition (no damage, loss, etc., apparent), and which does not bear such notation as "shipper's load and count."

clean collection (m) A collection in which a check or draft is present for payment without additional attached documentation.

clean credit (m) A letter of credit payable upon presentation of a draft, not requiring any other payment.

clean draft (m) A draft to which no documents have been attached.

clean on-board bill of lading (m) Shows cargo laden aboard a vessel with no exceptions as to condition or quantity.

clip on unit (m) A piece of refrigeration equipment which can be attached to an insulated (refrigerated) container or hung on the underside of a chassis to supply power to a container which does not have its own self-sustaining refrigeration unit.

clean transport document (m) Document which bears no superimposed clause or notation which expressly declares a defective condition of the goods and/or packaging.

clear record (m) A record that shows that the shipment was handled without any loss or damage being sustained.

clear record (t) A record which shows that a shipment was handled without loss or damage.

clearing house (t) An organization set up to process and collect bills for participating trucking companies.

collect charges (a) The transportation practice under which the receiver of the goods pays charges. *See charges collect.*

collecting bank (m) A bank that acts as an agent for the remitting bank. The collecting bank demands payment from the buyer and handles the funds received as instructed: generally, the funds are sent back to the remitting bank.

collection (m) A financial transaction in which a bank acts as an intermediary between two parties, (usually a buyer and seller located in different geographic areas), to facilitate the collection of a payment.

collect on delivery (COD) (a) A transportation service under which the purchase price of the goods is collected by the carrier from the receiver at the time of delivery, and subsequently, payment is transmitted by the carrier to the shipper. Carriers charge a nominal fee for this service. As the term COD implies payment is due upon delivery, there are no credit provisions in COD service. collect shipment (t) A shipment where collection of freight charges and advances are made by the delivering carrier.

collector of customs (m) A U.S. Treasury Department official in charge of Customs zone; collects duties, etc., on import shipments.

collector of customs (t) A representative of the United States Treasury Department acting for the government in connection with foreign traffic.

combi aircraft (a) Aircraft specially designed to carry unitized cargo loads on the upper deck of the aircraft forward of the passenger compartment.

combination aircraft (a) An aircraft capable of transporting both passengers and cargo on the same flight. Some cargo is carried on virtually all scheduled passenger flights in the belly pits below the passenger cabin.

combination rate (t) A rate made by combining two or more rates in different tariffs.

combined transport (m) Multiple types of transportation modes combined to make one continuous cargo flow.

commercial attaché (m) The commerce expert on the diplomatic staff of his or country's embassy.

commercial bill (m) A bill of exchange drawn by an exporter directly on the importer.

commercial invoice (m) Represents a complete record of the transaction between exporter and importer with regard to the goods sold. Also reports the content of the shipment and serves as the basis for all other documents about the shipment.

commercial vessel (m) Any merchant ship that is engaged in trading for cargo carrying purposes.

commodity (m) Article shipped. For dangerous and hazardous cargo, the correct commodity identification is critical.

commodity (t) Any article of commerce. Goods shipped.

commodity code (a) A system for identifying a given commodity by a number as a means of facilitating, for example, the application of computerization to freight transport. *See harmonized commodity description and coding system.*

common carrier (m) A transportation company operating under a certificate of convenience and necessity; provides service to the general public at published rates.

commodity exempt (t) One that may be transported in interstate commerce without operating authority or published rates.

commodity rate (t) A special rate on specific type of goods. A commodity rate replaces a class rate of the goods except when the tariff specifies the alternative use of class and commodity rates.

commodity tariff (t) A tariff containing only commodity rates.

common carrier (t) A transportation business that offers service to the general public. Interstate common carriers must hold a franchise issued by the Interstate Commerce Commission. this franchise limits service to a specific geographical area. Rates also are regulated. Routes and schedules of regular common carriers are regulated by government agencies, but irregular route common carriers may set their own without regulatory approval. Most states also regulate common carriers.

common carrier, irregular route (t) A common carrier whose routes and schedules are not regulated by government agencies.

Common Carrier Conference—Irregular Route (t) An independent and autonomous organization within the American Trucking Associations representing the carriers authorized as common carriers—irregular route.

common law (t) That system of law which does not rest for its authority upon any express statute, but derives its force and authority from consent, custom and usage. It is derived from law brought to the United States from England.

common ownership (t) Ownership of one mode of transportation by another. Such ownership is restricted by federal law.

common tariff (t) A tariff that gives the charges of two or more transportation lines.

complainant (t) One who makes a legal complaint; a plaintiff or prosecutor.

complaint (t) A statement of injury or grievance. *See formal complaint and informal complaint.*

computerized traffic flow (a) The increased use of electronic data processing is one of the most significant advances in physical distribution of recent years. Computers are being given new assignments not only by carriers, but also by shippers, to achieve better control over the flow of raw materials and finished inventories. Also the U.S. Government and the United Nations gather more accurate data on the flow of commerce. Airlines have been in the forefront of computer application of cargo operations by using computers to prepare air waybills, expedited terminal handling, pickup and delivery, up-to-the-minute shipment status, and billing.

concealed damage (m) Damage that is not evident from viewing the unopened package.

concealed damage (t) Damage to the contents of a package which is apparently in good condition externally.

concealed loss (m) Contents missing from a package that looks unopened.

concealed loss (t) Loss or damage that cannot be determined until the package is opened.

concurrence (tariff) (t) A document filed with the Interstate Commerce Commission by which a transportation lines agrees to be a party to a joint tariff.

conditions of contract (a) The terms and conditions established by the air carriers for the carriage of goods. These conditions are printed on the air waybill and include such items as limits of liability, claims limitations, indemnity and dimensional weight rules.

confirmed letter of credit (m) A letter of credit, issued by a foreign bank, the validity of which has been confirmed by a U.S. bank. An exporter whose payment terms are a confirmed letter of credit is assured of payment by the U.S. bank even if the foreign buyer or foreign bank defaults.

consign (t) To send or address goods to another.

consignee (a) A person named as the receiver of a shipment—one to whom a shipment is consigned.

consignee (m) A person or company to whom commodities are shipped. Officially, the legal owner of the cargo.

consignee (t) One to whom something is shipped.

consignee marks (t) Identification symbols placed on packages. They usually are a square, triangle, diamond, circle or cross with letters and/or numbers. They generally are used for export.

consignment (a) Shipment of one or more pieces of property, accepted by the carrier from one shipper at one time, receipted for in one lot, and moving on one air waybill.

consignment (m) 1. A stock of merchandise advanced to a dealer and located at his place of business, but with title remaining in the source of supply. 2. A shipment of goods to a consignee.

consignment (t) A shipment.

consignee account number (a) *See account number.*

consignor (a) One who designates the person to whom goods are to be sent. The consignor is usually the shipper.

consignor (m) A person or company shown on the bill of lading as the shipper.

consignor (t) The person by whom articles are shipped.

consolidator (a) An entity that provides service also provided by an air carrier, independent from that carrier, and derives income from package consolidation of other for tender to an air carrier. An air freight forwarder performs the functions of a consolidator.

consular declaration (a) A formal statement made to the consul of a foreign country describing goods to be shipped.

consular invoice (m) A document, required by some foreign countries, describing a shipment of goods and showing information such as the consignor, consignee, and value of the shipment. Certified by a consular official of the foreign country, it is used by the country's customs officials to verify the value, quantity, and nature of the shipment.

container (m) Any equipment used to unitize cargo, e.g., all types of containers and/or flats, whether International Standards Organization (ISO)—accepted or not, trailers, swap bodies, ro-ro equipment, igloos. Applies to all modes of transport.

container (t) Anything in which articles are packed.

container car (m) A specially designed rail car for transporting portable freight containers.

container freight station (CFS) (m) The physical facility where goods are received by carrier for loading into containers or unloading from containers and where carrier may assemble, hold, or store its container or trailers.

container load (m) A load sufficient is size to fill a container either by cubic measurement or by weight.

container manifest (m) Document showing contents of a loaded container.

container terminal (m) An area designated for the stowage of cargos in container; usually accessible by truck, railroad and marine transportation. Here containers are picked up, dropped off, maintained, and housed.

container (van body type) (t) A truck or trailer body provided with means for ready removal from and attachment to a vehicle.

container yard (CY) (m) A material-handling/storage facility used for completely unitized loads in containers and/or empty containers.

containerization (a) The practice or technique of using a boxlike device in which a number of packages re stored, protected, and handled as a single unit in transit. Container descriptions have been broadened to include a unitized load on a carrier owned pallet, loaded by shippers, and unloaded by receivers at places other an on airline premises and are restrained and contoured so as to permit proper positioning and tie-down aboard the aircraft. Containerized air freight reduces packaging costs for the shipper because of the protection afforded by the container. The buildup of container loads at the shipper's plant bypasses the terminals. Shippers that do not have appropriate facilities for loading and unloading containers may contract through the airlines for this service. Specially equipped trucks pick up containerized shipments at the shipper's plant for direct delivery to mechanical loading equipment near planeside. containerization saves the airlines time and staffing in ground handling and enables the carriers to achieve a more efficient use of the cubic capacity of modern freighter aircraft and wide body jets. Consequently the airlines pass along part of this savings to shippers.

containerization (t) Shipping system based on large cargo-carrying containers that easily can be interchanged between trucks, trains and ships without re-handling the contents.

container yard (CY) (m) Pont forest for a container. This can be a part of the terminal where vessels are loaded or can be an "off dock" facility which is not a part of the terminal loading area.

continuous seals (m) 1. Seals that remained intact on a truck from point of origin to destination. 2. Seals broken in transit by proper authority and without opportunity for loss before new seals were applied.

continuous seals (t) A term denoting that the seals on a truck remained intact during the movement of the truck from origin to destination; or, if broken in transit, that it was done by proper authority and without opportunity for loss to occur before new seals were applied.

contraband (m) Cargo that is prohibited.

contract carrier (t) A company that engages in for-hire transportation of property under individual contract or agreement with one or a limited number of shippers.

correction stamp (m) Used to make manual corrections to an original bill of lading.

correspondent bank (m) A bank that, in its own country, handles business of a foreign bank.

cost, insurance, and freight (m) Cost of goods, marine insurance, and all transportation (freight) charges are paid to the foreign point of delivery by the seller.

country of exportation (C/E) (a) Is usually, but not necessarily, the country in which merchandise was manufactured or produced and from which it was first exported. For example, merchandise made in Switzerland and shipped to the United States through Frankfurt, Germany, has as the country of exportation Switzerland.

courier (a) Attendant who accompanies shipment(s). Also, some courier companies provide a full transportation function, without accompanying attendants, offering door-to-door air service for time-sensitive documents or small packages on a same-day or next-day basis.

country risk (m) Financial risk base on the evaluation of political, economic, and social conditions of a country.

credit risk insurance (m) Protects the seller against loss due to default on the part of the buyer.

crew leasing (m) A term which probably describes the managerial function of agreeing to supply an entire "turnkey" crew for a given ship. This term would imply that the crewing management firm which gathered, and certified as competent such a crew, would probably pay them, while collecting its fee from the owner/operator. In this case, then, the crew would work for the crewing manager and not for the ship owner or charterer.

cubed out (m) Said of a ship when it is full by volume without being loaded down to its load line and not carrying the full dead weight tonnage to which it is entitled.

cubic capacity (t) The carrying capacity of a truck measured in cubic feet.

Cummins Amendment (t) An amendment to the Interstate Commerce Act relating to liability of transportation for loss of or damage to freight.

current ratio (t) Relationship of current liability to current assets.

customer automation (a) The use of air carrier automation equipment on the customer's premises that aids in the processing of shipments, i.e., airbill preparations, invoicing, weighing, and tracing.

customhouse (m) A government office where duties, etc., on foreign shipments are handled.

customhouse broker (a) A broker who is certified by the U.S. Bureau of Customs to ct for importers and other businessmen in handling for them the sequence of Customs formalities and other details attendant to the legal and expeditious importing of goods. Such brokers are among the specialists who have made it easier for more businessmen to participate successfully in international air commerce.

customs (a) A government authority designated to regulate flow of goods to/from a country and to collect duties levied by a country on imports and exports. The term also applies to the procedures involved in such collection.

customs (m) Government agency charged with enforcing the rules passed to protect the country's revenues.

customs bonded warehouse (m) A warehouse in any country authorized and established by Customs where goods may be stored.

customs court (a) A U.S. Customs Services court based in New York City consisting of three three-party divisions to which importers may appeal or protest classification and value decisions and certain other actions taken by U.S. Customs Service.

customs declaration (a) A statement, oral or written, attesting to the correctness of description, quantity, value, etc., of merchandise offered for importation into the United States.

customs entry (m) All countries require that the importer make a declaration on incoming foreign goods. The importer then normally pays a duty on the imported merchandise. The importer's statement is compared against the carrier's vessel manifest to ensure that all foreign goods are properly declared.

customs information exchange (CIE) (a) A clearinghouse of information for U.S. Customs Service officers.

customs invoice (m) A form requiring all data in a commercial invoice along with a certificate of value and/or a certificate of origin. Required in a few countries (usually former British territories) and usually serves as a seller's commercial invoice.

customs invoice (m) A document that contains a declaration by the seller, the shipper, or the agent of either as to the value of the goods covered.

customs tariff (m) A schedule of charges assessed by a government on imported or exported goods.

D&H (m) Abbreviation for dangerous and hazardous cargo. D&H materials require special handling both physically and in the system. See the hazardous materials manual for specific information on handling and booking D&H materials.

dangerous goods (a) Articles or substances which are capable of positing a significant risk to health, safety, or property when transported by air and which are classified according to the most current editions of the *ICAO Technical Instructions for the Safe Transport of Dangerous Goods by Air and the IATA.*

Dangerous Goods Regulations (a) Dangerous goods may be transported domestically and internationally by air. *Also see hazardous materials.*

dangerous goods (m) Term used by International Maritime Organization for hazardous materials.

date draft (m) A draft that matures in a specified number of days after the date that it is issued, without regard to the date of acceptance.

dead-head (t) A trip with an empty vehicle.

deadweight tonnage (DWT) (m) The total carrying capacity of a ship in metric tons. This figure includes the weight of fuel and stores as well as cargo. Current rules of thumb for estimating cargo capacity of a ship (cargo deadweight) are: 95 percent of deadweight for large tankers and 85 percent of deadweight for dry cargo/container ships. Smaller, older ships will have lower cargo deadweight as a result of less efficiency in design and higher bunker requirements to support less efficient engines. When military cargo is being carried it is normal for a ship to sail partially loaded, or for the ship to be full by volume before it reaches its limit in weight. The ship is then said to be cubed out. This is partly due to the nature of much military equipment which will not permit stacking, leading to waste of volume.

declaration (m) Statement from a ship on leaving port or passing a reporting station as to its next port of call. Ultimate ports of call can be given and intermediate ports omitted and the presence of a declaration is not, of itself, proof of where a ship intends to proceed. When used legitimately, the declaration that a ship is bound for a given port "for orders" or "awaiting orders" indicates that the ship is not carrying cargo beyond that port.

declared value for carriage (a) The value of goods declared to the carrier by the shipper for the purposes of determining charges or of establishing the limit of the carrier's liability for loss, damage, or delay. *See valuation charges.*

declared value for customs (a) The selling price of the contents or the replacement cost if the contents are not for resale. The amount must be equal to or greater than the declared value.

defendant (t) A person or party answerable to a legal complaint.

deferred air freight (a) Air freight requiring dependable, reliable service, but of a less time sensitive nature, with delivery provided over period of days.

delivering carrier (t) The transportation line that delivers a shipment.
delivery (t) The act of transferring possession of a shipment. This could be from consignor to carried, one carrier to another or carrier to consignee.

demise charter (m) *See bareboat charter.*

demurrage (a) The detention of containers by shippers or receivers of freight beyond a specified grace period. The airlines tender carrier owned containers to the customer for loading and unloading of the unit. In the event the container is not returned to the carrier within a specified time (usually 36–48 hours) a charge shall be assessed by the carrier for each 24-hour period or fraction thereof beyond the allowed time.

demurrage (t) Detention of a freight vehicle or container beyond a stipulated time. Also the payment made for such delay.

density (t) The weight of an article per cubic foot. The ratio of mass to bulk or volume.

deposition (t) Written testimony given under oath.

depot, container (m) Container freight station or a designated areas where empty containers can be picked up or dropped off.

destination (t) The place to which a shipment is to be delivered.

defferential route (t) Route for which there is no published rate. The rate must be computed from published rates on other routes.

dishonor (m) The refusal of a drawee to accept a draft or pay for it when due.

dismantled (m) Normally and accurately used to describe the removal of a ship's engines or otherwise having been rendered incapable of self-propelled navigation. Can also be less accurately used interchangeably with scrapped or broken up.

dispatch (m) 1. A foreign agent who sells directly for a supplier (seller) and maintains an inventory of the supplier's products. 2. To give an order to a drayman for the pickup or delivery of a specific container.

dispatching (t) The scheduling and control of intercity traffic and intracity pickup and delivery.

disposal (m) Normally the deletion from a fleet or group of ships (such as a company) of an individual ship or group of ships. Ships sold to be broken up or sold to another registered owner/operator would be referred to as disposed of by the losing firm. Ships for which charter arrangements are terminated will leave the fleet of a given operator or a service. They would not normally be referred to as having been disposed of, however. A ship may be listed by an owner as "for disposal" when it has been designated as being for sale. The ship can continue to operate or can be laid up, at the owner/operator's discretion, while awaiting the disposal. Ships listed for disposal for scrapping often continue to exist for long periods of time without operating while the owner waits for scrap prices and other conditions to be favorable.

distance rate (t) Charge made on basis of miles traveled, often used as one component of a freight rate.

distribution (t) Generally considered to be the act of delivering less-than-truckload shipments within a city or an area beyond.

diversion (t) A change made in the route of a shipment in transit.

divert (m) The route of a shipment changed in transit from that shown on the original billing. Used interchangeably with re-consign.

divert (t) To change the route of a shipment in transit.

division (t) The apportionment by carriers of revenue received from joint traffic.

dock (t) A platform where trucks load and unload.

dock receipt (t) A receipt given for a shipment received or delivered at a pier or dock. When delivery of a foreign shipment is completed, the dock receipt is surrendered to the transportation line and a bill of lading is issued.

document holder (m) A receptacle on the container door or front wall, sealed and containing the certificate of container approval.

documentary collection (m) A collection in which a draft is accompanied by shipping or other documents.

documentary credit (m) A letter of credit that requires documents to accompany that draft of demand for payment.

documentary draft (m) A draft to which documents are attached.

documents (m) Items presented along with the draft. These may include the bill of lading, the commercial invoice, the marine insurance policy or certificate of origin, the weight list, the packing list, and the inspection certificate (or certificate of analysis). Under a letter of credit, the documents must be examined for compliance before the draft is paid.

documents against acceptance (D/A) (m) Instructions given by a shipper to a bank indicating that documents transferring title to goods should be delivered to the buyers only upon the buyer's acceptance of the attached draft.

documents against payment (D/P) (m) An indication on a draft that the documents attached are to be released to the drawee only on payment.

draft (bill of exchange) (m) An instrument, much like an ordinary check in appearance, which is used as a formal demand for payment in a business transaction.

drawee (m) The individual or firm on whom a draft is drawn and who owes the stated amount.

drawer The issuer or signer of a draft.

drawing (m) Presentation of the draft and documents required by the terms of a letter of credit.

drayage (t) The charge made for hauling freight on carts, drays or trucks.

dropoff (a) Shipper delivers shipment(s) to air carrier for transportation.

due-bill (t) Bill rendered by carrier for undercharges.

dunnage (t) The material used to protect or support freight in trucks. The weight of dunnage is shown separately on the bill of lading since it is material used around a cargo to prevent damage. Often it is transported without charge.

duty (a) A tax imposed on imports by the Customs authority of a country. Duties are generally based on the value of the goods (ad valorem duties). Some other factor such as weight or quantity (specific duties) or a combination of value and other factors (compound duties).

duty (t) A tax levied by a government on imports and exports.

dynamic seal *See active seal.*

E-cup seal This seal consists of two small metal cups. When snapped together, they form a closed chamber which protects the knot or crimp on a piece of string or wire passing through the closure hasp on the container.

EDI (m) Electronic data interchange.

electronic seal A seal that contains electronics and is electrically powered, often by batteries. (If the device uses active fiber optics, it is usually classified under that category.) Electronic seals typically check continuously for signs of tampering. If they report tampering immediately (in "real-time"), they may be considered intrusion or burglar alarms, rather than seals.

Elkins Act (t) Law providing penalties for giving rebates and other violations of Interstate Commerce Act.

embargo (a) Temporary refusal to accept traffic for transportation at certain points or in certain routes due to emergencies, limitations, of facilities, or other abnormal circumstances.

embargo (t) To resist or prohibit the acceptance and handling of freight. A formal notice that certain freight will not be accepted.

endorsement (m) A signature on the back of a negotiable instrument made primarily for the purpose of transferring the rights of the holder to some other party. It is a contract between the holder and all parties of the instrument.

enroute (t) On the way.

entry (customs) (t) A statement of the kinds, quantities and values of goods imported together with duties due, if any, declared before a customs office or designated officer.

equipment interchange receipt (EIR) (m) A form used by the parties delivering or receiving containers and container equipment. It is used for equipment control and damage purposes. Synonymous with trailer interchange receipt (TIR).

estimated weight (t) The weight specifically stated in tariffs for goods shipped in specified packages, or in a specific manner.

examiner (ICC) (t) A representative of the Interstate Commerce Commission vested with power to administer oaths, examine witnesses, take testimony, and otherwise conduct hearings of cases submitted to or initiated by the Commission.

exception ratings (a) Rates set at a certain percentage above the general commodity rates because they apply to commodities that require special handling, such as live animals, human remains, or automotive vehicles.

exceptions to classification (t) A publication containing classification ratings and rules different from the classification ratings and rules contained in a major classification.

exchange bill of lading (t) A bill of lading issued in exchange for another bill of lading.

exhibit (t) A document or physical object introduced in evidence.

expediting (t) To accelerate a process. Expedited freight service is usually superior to normal service. Dispatching less than truckload quantities on a single truck for quick delivery is an example of expedited service. Such service frequently necessitates payment of "exclusive use of vehicle" freight charges.

expiration notice (t) A notice in a tariff that all or some part of the tariff will expire at a stated time.

export broker (m) An individual or firm that brings together buyers and sellers for a fee, but does not take part in actual sales transactions.

export control (a) To exercise control over exports for statistical and strategic purposes, Customs enforces the export control laws for the U.S. Department of Commerce and other federal agencies.

factor (m) A factor is an agent who will, at a discount (usually 5 percent to 8 percent of the gross), buy receivables. This agent can be either a bank or a private specialist. The factor makes a profit on the collection and provides a source of cash flow for the seller, although less than if the business had held out to make the collection itself.

FAK (m) Freight all kind.

false billing (t) Describing freight on shipping documents so as to misrepresent the actual contents.

false billing (m) Misrepresenting freight or weight on shipping documents.

feeder service (t) Short transportation lines running from a truck line into nearby areas to collect and distribute freight for the main line. Usually 25 to 35 miles long.

fiber optic seal A seal that uses fiber optics. The cable is an optical fiber or bundle of optical fibers. Cutting the optical fibers changes their light transmission or other properties. Fiber optics can be used to make both passive seals and active seals. For an active seal, light pulses re sent down the optical fiber(s) many times per second. If the optical fibers are cut, the light pulses fail to complete the loop and this is detected by electronics.

finance docket (t) An Interstate Commerce Commission list of scheduled cases relating to the financing, extension, abandonments and consolidations of common carriers.

fixed charges (t) Carrier costs that do not vary with an increase or decrease in traffic. An accounting classification. Sometimes called fixed operating costs.

flag of convenience register (m) A national register offering registration to a merchant ship not owned in the flag state. The major flags of convenience (FOC) attract ships to their register by virtue of low fees, low or non-existent taxation of profits, and liberal manning requirements. True FOC registers are characterized by having relatively few of the ships registered actually owned in the flag state. Thus, while virtually any flag can be used for ships under a given set of circumstances, a FOC register is one where the majority of the merchant fleet is owned abroad. Also referred to as an open register.

flag state (m) The nation in which a ship is registered and which holds legal jurisdiction as regards to operation of the ship, whether at home or abroad. Differences in Flag State maritime legislation determines how a ship is manned and taxed, and whether a foreign owned ship may be placed on the register.

flammable liquids (m) Liquids that give off vapors that are combustible at a certain temperature. These are sometimes called inflammable but flammable is preferred.

flammable liquids (t) Liquids that give off vapors capable of burning.

FOB (m) *See free on board.* Shipped under a rate that includes costs of delivery to and the loading onto a carrier at a specific point.

force majeure (m) The title of a common clause in contracts, typically exempting the parties for non-fulfillment of their obligations as a result of conditions beyond their control, such as earthquakes, floods, or war.

foreign draft (m) A draft drawn by a bank on a foreign correspondent bank and sent to the payee.

foreign trade zone (a) A port designated by the Government of a country for duty free entry of any non-prohibited goods. Merchandise may be stored, displayed, used for manufacturing, etc., within the zone and re-exported without duties being paid. Duties are imposed on the merchandise (or items manufactured from the merchandise) only when the goods pass from zone into an area of the country subject to the Customs authority.

foreign trade zone (m) A free port in a country divorced from Customs authority but under government control. Merchandise, except that which is prohibited, may be stored in the one without being subject to U.S. tariff regulations.

foreign trade zone entry (m) A form declaring goods which are brought duty free into a foreign trade zone for further processing or storage and subsequent exportation.

formal complaint (before ICC) (t) A complaint filed with the Interstate Commerce Commission for investigation and adjudication under formal procedures provided by Commission rules. *See informal complaint.*

forty-foot equivalent unit (FEU) (m) Term indicating capacity of container vessel or terminal.

foul bill of lading (m) A receipt for goods issued by a carrier with an indication that the goods were damaged when received.

free alongside (FAS) (m) A shipping term under which the seller quotes a price including delivery of the goods alongside the vessel and within reach of its loading tackle.

free-astray (t) A shipment miscarried or unloaded at the wrong terminal is billed and forwarded to the correct terminal free of charge because it is astray.

freeboard (m) Vertical measurement taken amidships from the load waterline, or loadline, to the underside of the freeboard deck.

freeboard deck (m) Uppermost watertight deck or deck having weather openings capable of being hermetically sealed.

free on board (FOB) (t) Usually indicates place where responsibility for expenses and risk for goods is passed from seller to buyer. For example, FOB motor carrier would usually mean that a price quoted for goods would include loading on a truck at the seller's building. This term is not always used precisely, and it is best to qualify it to show exactly what is covered.

free on board (U.S. domestic use) (m) Shipped under a rate that includes costs of delivery to and the loading onto a carrier at a specific point.

FOB freight allowed: The same as FOB names inland carrier, except the buyer pays the transportation charge of the inland carrier.

FOB freight prepaid: The same as FOB named inland carrier, except the seller pays the freight charges of the inland carrier.

FOB named point of exportation: Seller is responsible for the cost of placing the goods at a named point of exportation. some European buyers use this form when they actually mean FOB vessel.

FOB vessel: Seller is responsible for goods and preparation of export documentation until actually placed aboard the vessel.

free port (m) A restricted area at a seaport for the handling of duty-exempted import goods. Also called a foreign trade zone.

free time (t) The period freight will be held before storage charges are applied.

free trade zone (a) *See foreign trade zone.*

free trade zone (m) A port designated by the government of a country for duty-free entry of any non-prohibited goods. Merchandise may be stored, displayed, used for manufacturing, etc, within the zone and re-exported without duties.

freight (a) Property, commodities of all kinds, including small package service tendered to an airline for transportation. Does not include mail, express, or passenger baggage.

freight (t) Anything being transported.

freight (m) Refers to either the cargo carried or the charges assessed for carriage of the cargo.

freight bill (t) Document for a common carrier shipment. Gives a description of the freight, its weight, amount of charges, the rate for charges, taxes and whether collect or prepaid. If the charges are paid in advance or are to be collected at the origin, it is called a prepaid freight bill. If charges are to be collected at destination, it is called a destination or collect freight bill.

freight bill (m) A document issued by the carrier based on the bill of lading and other information; used to account for a shipment operationally, statistically, and financially.

freight charge (t) Payment due for freight transportation.

freight charge (m) The charge assessed for transporting freight.

freight claim (t) *See claim.*

freight claim (m) Demand upon a carrier for the payment of overcharge or loss or damage sustained by a shipper or consignee.

freight classification (t) *See classification rating.*

freight correction notice (m) A document used to record any and all cost changes to the freight manifest (corrected B/L). These changes may be additions, deletions or corrections of information affecting charges.

freighter (a) *See all-cargo aircraft.*

freight forwarder (t) A company that assembles small shipments from various shippers into large shipments, usually full truck or car load. Forwarders send these consolidated shipments to a station where they are disassembled and routed to the proper destinations. These stations are called break bulk stations. Forwarders try to make up a full truck or car load shipment to take advantage of the better rate for a full load.

freight forwarder (m) A carrier that collects small shipments from shipper, consolidates the small shipments and uses a basic mode to transport these consolidated shipments to a destination where the freight forwarder delivers the shipment to the consignee. A person whose business is to act as an agent on behalf of the shipper. A freight forwarder frequently makes the booking reservation. An independent business that handles export shipments for compensation.

full container load (FCL) (m) The maximum permissible weight for the value of the cargo carried in a container.

fwd (m) Abbreviation for forward.

gateway (t) A point where freight moving from one territory to another is interchanged between transportation lines.

general commodity rate (a) An air freight rate applicable on all commodities except those for which specific rates have been filed. Such rates are based on weight and distance and are published for each pair of cities an airline serves.

general order (GO) (a) Merchandise not entered within five working days after arrival of the carrier and then stored at the risk and expense of the importer.

goods (t) Merchandise.

grandfather rights (t) The right provided in the Motor Carrier Act for a motor common carrier to obtain a certificate of public convenience and necessity to operate over the route or routes over which it or its predecessor in interest was in bona fide operation on June 1, 1935, without further proof of public convenience and necessity. The right of a contract carrier by motor vehicle to a Permit as authority to operate over the route or routes over which it or its predecessor in interest was in bona fide operation on July 1, 1935, without further proof. This term also may be applied to other forms of transportation.

gray area (t) Except for those carrying certain exempt commodities, for-hire trucking companies must be certificated and are regulated by state and federal governments. "Gray area" carriers illegally operate as for-hire carriers through the use of some form of subterfuge or camouflage.

gross tonnage (GRT) (m) A statutory measure of permanently enclosed volume in a merchant ship. Originally 100 cubic feet of space equaled one gross ton, and there has never been a way to convert GRT to weight of ship or cargo. A 1969 convention resulted in international agreement on a formula yielding a pure number that relates neither to space nor to weight. The convention also eliminated a situation in which differences in national rules resulted in sometimes radical changes in a ship's GRT when it changed flag. the GRT, and numbers derived from it, such as net registered tonnage (NRT), form the basis for national registration fees and port and canal dues. While it gives a good approximation of relative size of two ships of the same type, it is a misleading measure of carrying capacity. When cited in reports, GRT is most useful as a means of differentiating a ship from other possible ships of the same name.

gypsy (t) An independent truck operator who drives his own truck and obtains freight wherever he can.

harmonized commodity description and coding system (a) A multipurpose international goods classification system designed to be used by manufacturers, transporters, exporters, importers, Customs, statisticians, and others in classifying goods moving in international trade under a single commodity code. Developed under the auspices of the Customs Cooperation Council (CCC), an international Customs organization in Brussels, this code is a hierarchically structured product nomenclature containing approximately 5,000 headings and subheadings describing the articles moving in international trade. It is organized into 99 chapters arranged in 22 sections with the sections generally covering an industry (e.g., Section XI, Textiles and Textile Articles) and the chapters covering the various materials and products of the industry (e.g., Chapter 50—Silk; Chapter 55—Manmade Staple Fibers; Chapter 57—Carpets). The basic code contains 4 digit headings and 6 digit subheadings. The United States will add digits for tariff and statistical purposes. In the United States duty rtes will be the 8-digit level; statistical suffixes will be at the 10-digit level. This Harmonized System (HS) is scheduled to supplant the current U.S. Tariff Schedule (TSUSA) in January 1988.

hazardous materials (a) Means a substance or material which has been determined by the U.S. Secretary of Transportation to be capable of posing an unreasonable risk to health, safety, and property when transported in commerce and which has been so designated.

Title 49, Code of Federal Regulations (U.S.) Transportation—Parts 100–199, govern the transportation of hazardous materials. Hazardous materials may be transported domestically, but they may be classified as dangerous goods when transported internationally by air. *Also see restricted articles and dangerous goods.*

haz mat (m) An industry abbreviation for hazardous material.

hearing (t) A legal proceeding for formal presentation of evidence. In the trucking industry, usually refers to a case before the Interstate Commerce Commission or a state regulatory agency.

high-value cargo A shipment containing one or more of the following commodities; artwork, money, bills of exchange, pearls, bonds, platinum, bullion, precious metal, computes, negotiable securities, computer componetry, coined or un-coined silver bullion, all related technical material, stamps, postage or revenue, currency, stock certificates, finished furs, watches and watch movements, cut or uncut gems (excluding watch parts), coined or un-coined gold bullion, jewelry (other than costume jewelry).

hog, hogging (m) A ship is said to be hogged when the ends of the ship are depressed below the center. This may result from improper cargo stowage, with too much weight at the ends. It is a naturally occurring tendency when a ship is suspended at the crest of a wave with each end relatively unsupported. May cause strain and buckling of hull plates, and in extreme cases, breaking in two. A ship without sheer will give the impression of being hogged.

hold for pickup (a) Freight to be held at the carrier's destination location for pickup by the recipient.

house carrier (m) A trucking company that has a contractual relationship with a carrier.

hundredweight pricing (a) Special pricing for multiple-piece shipments traveling to one destination which are rated on the total weight of the shipment (usually over 100 pounds) as opposed to rating on a per package basis.

hybrid seal *See barrier seal.*

idle (m) A merchant ship may be idle between periods of employment carrying cargo. Nevertheless, the ship will be fully provisioned and manned, as opposed to when it is in a layup status.

immediate delivery (a) A procedure allowing more rapid release of importers of imports flown into the United States each year. Under the immediate delivery program, an importer can take delivery almost immediately of virtually all of an incoming shipment. Customs retains a sample for the purposes of appraisal. On the basis of this appraisal the importer is billed later for payment of duties.

IMO number (m) A seven-digit number allotted to every merchant ship hull which is its internationally recognized identity number, introduced by International Maritime Organization (IMO) resolution to assist registration and combat fraud. For ships of all flags and registries this is the same as the Lloyd's Register Number or LRN.

import (t) To bring in goods from a foreign country.

import license (a) A document required and issued by some national governments authorizing the importation of goods into their individual countries.

in bond (a) As applied to air freight coming into the United States, the term "In Bond" designates a procedure under U.S. Customs clearance of cargo is postponed until the cargo reaches an inland Customs point rather than subjecting the cargo to clearance procedures at the first arriving United States gateway airport where the process might be more time consuming. The procedure is so named because the cargo moves under the carrier's bond (financial liability assured by the carrier) from the gateway airport and remains "in bond" until Customs releases the cargo at the inland Customs point (airport.)

in bond (m) Cargo moving under Customs control where duty has not yet been paid.

in bond (t) Goods on which a duty or tax is due are "in bond" when placed in the custody of a government or bonded warehouse or are traveling by a bonded carrier. Bonding guarantees that the duty will be paid.

inbound (m) Cargo moving from a rail terminal toward its destination. Generally used for cargo coming off a train and heading for final delivery to consignee.

indemnity bond (m) An agreement to hold a carrier harmless with regard to a liability.

indemnity bond (t) *See bond of indemnity.*

indicating seal This term is sometimes used to distinguish the device from a barrier seal.

indirect air carrier (a) *See air freight forwarder.*

individual tariff (t) A tariff issued by a transportation line individually.

inflammable liquids (t) *See flammable liquids.*

informal complaint (before ICC) (t) 1. A complaint filed informally with the Interstate Commerce Commission to obtain its permission to remedy a condition when the complainant and defendant are in agreement as to the merits of the complaint. 2. A complaint filed informally with the Interstate Commerce Commission in a case which may be handled through correspondence with interested parties. *See formal complaint.*

informal entry (a) A simplified import entry procedure accepted at the option of Customs for an noncommercial shipment (baggage) and any commercial shipment not over $1,000 in value.

inherent advantage (t) A natural or innate advantage. For instance, inherent advantages of motor carriers over other modes of carriers include flexibility, ability to run through city streets and to make delivery of goods and passengers at more convenient points in centers of congestion; relatively low first cost of initiating operations; and ability because of its small units of transportation, to furnish more frequent service. The Interstate Commerce Act instructs the ICC to "recognize and preserve the inherent advantages" of each mode of transportation.

inherent vice (m) An insurance term referring to any defect or other characteristic of a product that could result in damage to the product without external cause (for example, instability in chemical that could cause it to explode spontaneously). Insurance policies may specifically exclude losses caused by inherent vice.

initial carrier (m) The transportation line to which a shipment is delivered by the shipper.

initial carrier (t) The transportation line that picks up a shipment from the shipper.

initial point (t) The point at which a shipment originates.

inland bill of lading (m) A bill of lading used in transporting goods overland to the exporter's or from the importer's international carrier. A through bill of lading is many times substituted for an inland one.

inland carrier (m) A transportation line that hauls export or import traffic between ports and inland points.

inland carrier (t) A transportation line which hauls export or import traffic between ports and inland points.

inspection certificate (m) A certificate issued by an independent agent or firm attesting to the quality and/or quantity of the merchandise being shipped. Such a certificate is usually required in a letter of credit for commodity shipments.

insurance (m) An insurance policy or certificate normally covers the shipments of merchandise from the time they leave the warehouse at the shipping point until they reach the destination point named in the policy or certificate.

insurance, all-risk (m) This type of insurance offers the shipper the broadest coverage available, covering against all losses that may occur in transit.

insurance certificate (m) Where the seller provides ocean marine insurance, it is necessary to furnish insurance certificates, usually in duplicate. The certificates are negotiable documents and must be endorsed before submitting them to the bank. The seller can arrange to obtain an open cargo policy that the freight forwarder maintains.

insurance, freight (t) Insurance that covers one or more perils that goods may encounter in transportation.

insurance, general-average (m) This clause covers damages or losses arising from maritime ventures or losses arising from maritime ventures for all parties shipping or owing.

integrated cargo service (a) A blend of all segments of the cargo system providing the combined services of air carrier, forwarder, ground handling and agents.

interchange (t) Transfer of freight from one carrier to another.

interchange points (t) A terminal where freight is transferred from one transportation line to another.

interline (a) The movement of a shipment via two or more carriers. *See intermodal compatibility.*

interline (t) Between two or more transportation lines.

interline freight (m) Freight moving from point of origin to destination over the lines of two or more transportation lines.

interline freight (t) Freight moving from point of origin to destination over the lines of two or more transportation companies.

interline waybill (m) A waybill covering the movement of freight over two or more transportation lines.

interlocking directorate (t) Influence in the management of two or more outwardly competitive companies by making a person a director in those companies. The Interstate Commerce Act makes it illegal for any person to be a director of more than one common carrier without special authorization.

intermediate carrier (m) A transportation line that moves cargo on a route that does not include the point of origin or destination.

intermediate carrier (t) A transportation line hauling a shipment between the originating and the delivering carrier.

intermediate clause (t) A section of a tariff giving rtes to a point not named in the tariff, but which is between points that are named.

intermediate point (t) A point located between two other specific points.

intermodal (m) Used to denote movement of large cargo containers interchangeably between motor, water, and air carriers and where the equipment is compatible with the multiple systems used.

intermodal compatibility (a) The capability which enables a shipment to be transferred from one form of transport to another, as from airplane to highway truck, to railway freight car, to ocean vessel. Some aircraft in service today have the capability for intermodal exchange of the large types of standard containers currently used in surface transport.

intermodal transport (m) Moving ocean freight containers by various transportation modes. The fact that the containers are of the same size and have common handling characteristics permits them to be transferred from truck to railroad to air carrier to ocean carrier.

internal register (m) A register of ships maintained as a subset of a national register. Ships on the internal register fly the national flag and have that nationality, but are subject to a separate set of maritime rules from those on the main national register. These differences may include lower taxation of profits, manning by foreign nationals, and ownership outside the flag state (when it functions as a flag of convenience register). The Norwegian International Ship Register and Danish International Ship Register are the most notable examples of an internal register. Both have been instrumental in stemming flight from the national flag to flags of convenience and, to an extent, in attracting foreign owned ships to the Norwegian and Danish flags.

Interstate (t) Between states.

interstate commerce (m) Commerce from: (1) one state, territory or district of the United States, to any other: (2) from any place in the United States through a foreign country to any other place in the United States; or (3) from a foreign country but only insofar as such transportation takes place within the Unites States.

interstate commerce (t) The Interstate Commerce Act defines interstate commerce as transportation from one state or territory of the United States or the District of Columbia to any other, or from any place in the United States through a Foreign Country to any other place in the United States, or from or to any place in the United States to or from a foreign country, but only insofar as such transportation takes place within the United States.

Interstate Commerce Act (t) An act of Congress regulating the practices, rates and rules of transportation lines engaged in handling interstate traffic.

Interstate Commerce Commission (ICC) (t) The federal body charged with enforcing acts of Congress affecting interstate commerce.

interstate traffic (t) *See interstate commerce.*

in transit (m) In transit, or in passage.

investigation and suspension docket (t) List of cases before the Interstate Commerce Commission which deal with suspension or tariff, rates, or rules.

IPI (m) Abbreviation for Inland Point Intermodal which refers to inland points (non-ports) that an be served by carriers in a through bill of lading.

irregular common carrier (t) See common carrier—irregular route.

irrevocable letter of credit (m) Letter of credit in which the specified payment is guaranteed by the bank if all terms and conditions are met by the drawee and which cannot be revoked without joint agreement of both the buyer and the seller.

issuing carrier (t) A transportation line that publishes a tariff or issues a bill of lading or other document.

JIT (just in time) (a) The principle of production and inventory control that prescribes precise controls for the movement of raw materials, component parts, and work-in-progress. Goods arrive when needed (just in time) for production or use rather than becoming expensive inventory that occupies costly warehouse space.

joint boards (t) Boards consisting of members from several states may be created by the Interstate Commerce Commission (under provisions of Section 205 of Motor Carrier Act) to hear cases.

joint rate (a) A single through-rate on cargo moving via two or more air carriers on air and surface carriers.

joint rate (t) A rate for hauling a single shipment over two or more independent transportation lines, which cooperate to offer a through service. The shipment travels as one bill of lading.

joint routes (t) Routes established by two or more connecting carriers for the continuous through movement of traffic over their lines.

joint through rate (t) *See joint rate.*

joint traffic (t) Traffic moving between stations located on one transportation line and stations located on another transportation line.

joint venture (m) A venture or business activity undertaken by two or more people or firms in a merger or partnership. For purposes of maritime joint ventures, the individual ships participating need not be jointly owned, but rather could be contributed to a joint service. Management and operations of ships in a joint venture could be by the original owner(s), by the joint venture company, or by a third party manager hired for this purpose. None of these arrangements would affect ownership, per se, and would not therefore be subject to our routine scrutiny.

jurisdiction (t) The power of a court or other tribunal to consider and decide a question. For example, the Interstate Commerce Commission has jurisdiction over rates, rules and regulations.

knocked down (t) A term denoting that an article is partially or entirely taken apart (not set up).

known damage (m) (t) Damage discovered before or at the time of delivery of a shipment.

known loss (m) (t) A loss discovered before or at the time of delivery of a shipment.

lading (t) That which constitutes a load. The freight in a vehicle.

lay order (a) The period during which the imported merchandise may remain at the place of unlading without some action being taken for its disposition, i.e., beyond the 5-day General Order period.

layup (m) An idle merchant ship will be placed in layup when no immediate employment seems in prospect. Layup will normally involve partial preservation of engineering spaces and removal of all or most of the crew. A laid up ship cannot, therefore, be immediately reactivated. This is different from the case of a ship being dismantled which renders the ship incapable of reactivation without installation of propulsion equipment.

lead seal A small piece of lead with holes for the passage of the sealing wire or string. The wire (or string) is passed through the closure hasp on the container, and then through the holes the lead, which is then compressed to embed the wire. Often a logo or serial number is embossed into the lead during compression. Sometimes a soft, non-lead alloy is used in place of lead; such a seal may still be (misleadingly) called a lead seal or a lead-wire seal.

lead-wire seal *See lead seal.*

lease (m) Not correctly used as the equivalent of charter. A ship may be leased to an operator by the parent company which is not a ship operator. For example, banks frequently become the owners of ships, either intentionally as part of a speculative new building, or unintentionally as part of a repossession. The effect of a lease arrangement will be essentially the same as that of a bareboat charter.

letter of credit (l/c) (a) A document issued by a bank per instructions by a buyer of goods authorizing the seller to draw a specified sum of money under specified terms, usually the receipt by the bank of certain shipping documents, within a given time.

letter of indemnity (m) When a bank does not wish to accept a foul bill of lading, the shipper (so that there shall be nothing to prevent the draft in question from being discounted, or for any other reason) may obtain a clean bill of lading. To do so, the shipper signs a letter of indemnity on the basis of which may be obtained the clean bill of lading, although the dock or mate's receipt showed that the shipment was damaged or in bad condition.

lien (m) (t) A legal claim on goods for the satisfaction of some debt or duty.

lightering (m) The discharge or, more rarely, loading of cargo while at anchor. This is done where port facilities are inadequate, where a port is overcrowded, or where several small cargoes are being consolidated into one. This operation is most common with tankers or with bulk carriers carrying grain, and least common with container ships and roll-on/off ships.

line (m) Although often used interchangeably with ship owner, the term line correctly refers to a service, usually of regularly scheduled ships over an advertised route. The ships need not, and often do not belong to a single owner.

line haul (t) Movement of freight between cities and terminals. Line haul does not include pickup and delivery service.

line of credit (m) A commitment by a bank to a borrower to extend a maximum amount of credit in a series of transactions. A line of credit, acceptance and/or discounting of drafts, multiple loans of drafts, multiple loans of advances and other forms of credit.

liquidate (t) To settle the accounts and distribute the assets of a business which is ceasing operations.

Lloyd's registry number (LRN) (m) *See IMO number.*

load line (m) The mark permanently inscribed and painted on the side of a ship, amidships, which designates the greatest depth to which it can be legally loaded. The mark consists of a circle bisected by a horizontal bar. When the circle is half submerged the ship is at its full deadweight capacity. Also known as the Plimsoll Mark. The painted boot topping of a merchant ship hull does not have to stop at the load waterline and may continue beyond or stop short. The only reliable indicator of whether a ship is loaded or in ballast is the relation of the load line to the waterline.

local rate (t) A rate applying between stations located on the same transportation line.

local tariff (t) A tariff containing rates applicable only between terminals located on the same transportation line.

lock Hardware that resists, complicates, and/or delays unauthorized entry or removal of items.

locking bar (m) Device that secures container doors at top and bottom.

log book (t) A book carried by truck drivers. They contain daily records of hours, routes, etc. They are required by Interstate Commerce Commission regulations.

longshoreman (m) Individual employed locally in a port to load and unload ships.

lot labels (a) Labels attached to each piece of a multiple lot shipment for identification purposes.

lower deck containers (a) Carrier owned containers specifically designed as an integral part of the aircraft to fit in the cargo compartments (lower deck) of the wide body aircraft.

main deck (m) The uppermost continuous deck of a ship, running fore to aft. Normally, the freeboard deck.

manager (m) A company which specifically manages (crews, operates, finds cargo for, insures, performs accounting functions for, etc.)

a ship or group of ships. May also, in fact own the ships, either as a registered owner or as a parent company (managing owner). The manager and charterer, if any, exercise more day-to-day operational control over a ship than does the registered owner or parent company, under normal circumstances.

managing owner (m) One, usually of several co-owners, to whom the others have delegated management of the ship or ships. The term managing owner at least implies that the ship is owned by more than one firm or set of partners.

manifest (m) A full list of a ship's cargo, extracted from bills of lading. A copy, known as the outward manifest, is lodged with the customs authorities at the port of loading. Another, the inward manifest, is lodged at the discharge port, with a copy for the ship's agent (now frequently transmitted electronically) so that unloading may be planned in advance.

manifest (t) A document describing a shipment or the contents of a vehicle or ship.

manifest miles (t) Miles traveled in dispatched revenue service between terminals.

marine insurance (m) Broadly, insurance covering loss or damage of goods at sea. Marine insurance typically compensates the owner of merchandise for losses sustained from fire, shipwreck, piracy, and various other causes but excludes losses that can be legally recovered from the carrier.

marks (a) Information placed on outer surface of shipping containers or packages such as address labels, box specifications, caution, or directional warnings.

marks (t) Letters, numbers, and characters put on a package for identification.

maximum rate (t) The highest lawful rate that may be charged.

memorandum bill of lading (t) A duplicate copy of a bill of lading.

memorandum tariff (a) Publications which contain rule and rate information extracted from official tariffs. Memorandum tariffs are published by many carriers and are available from these carriers upon request.

merchant marine (m) All ships engaged in the carriage of goods. All commercial vessels (as opposed to all non-military ships) which excludes tugs, fishing vessels, offshore oil, etc. Also, a grouping of merchant ships by nationality or register.

merchant ship (m) A vessel that carries goods against payment of freight. Commonly used to denote any non-military ship, but accurately restricted to commercial vessels, only.

metal cable seal *See cable seal.*

metal ribbon seal A seal made from sheet metal. One end of the ribbon snaps irreversibly into a head on the other end. Popular for use on railcars. Moe robust than a plastic strap seal, though still not considered a barrier seal. Also known as a car-box or car-ball seal.

mileage tariff (t) A tariff containing rates applied according to distance.

minimum charge (a) The lowest rate applicable on each type of air cargo service no matter how small the shipment.

minimum charge (t) The least charge for which a shipment will be handled.

minimum rate (t) The lowest lawful rate that may be charged.

mixing privileges (t) Provides for the mixing of truckload freight according to various combinations and alternatives. The rate is calculated as if each of the articles comprising the mixed truckload were shipped as truck loads.

mode (t) Frequently used to refer to the basic divisions of the transportation industry. The principal modes of transportation are truck, air, and water.

motor carrier (m) A company that moves trailers via highways.

Motor Carrier Act, 1935 (t) An Act of Congress providing regulation of motor carriers. Now Part II of the Interstate Commerce Act.

motor truck (t) Single, self-propelled motor vehicle carrying its load on its own wheels. Primarily designed for the transportation of property.

motor vehicle (t) Any vehicle, self-propelled or drawn by mechanical power, designed for operation on the highways or natural terrain in the transportation of property or passengers.

munitions carrier (t) A company that transports munitions by motor vehicle.

nationality (m) For a ship this is provided by the flag flown as a result of having been legally entered onto the register of a country which maintains such a shipping register. All legal jurisdictions over the ship resides with the flag state which grants nationality to the ship, regardless of the nationality or location of the vessel's actual owner.

negligence (t) Failure to exercise the degree of care the law demands.

negotiable instrument (m) Any written evidence of an obligation which may be transferred by Endorsement or by delivery, and of which the transferee may become a holder in due course. Examples include: checks, bills of exchange, drafts, promissory notes and some types of bonds or securities.

negotiable letter of credit (m) The letter of credit which allows the beneficiary to present the documents and draft to a named bank (the nominated bank) or to any bank willing to negotiate them on or before the expiration date.

negotiate (letter of credit) (m) Giving of value for draft(s) and/or document(s) by the bank authorized to negotiate.

NES (a) The abbreviation (not elsewhere specified) often appears in air freight tariffs. Fro example: "advertising matter, NES; printed matter, NES" indicating that the rate stated in the tariff applies to all commodities within the commodity group except those appearing under their own rate. The abbreviation NES as used in air freight tariffs, is comparable to the abbreviation NOIBN (not otherwise indexed by number) and NOS (not otherwise specified) which appear in tariffs published by the surface modes.

neutral air waybill (a) A standard air waybill without identification of issuing carrier.

non-vessel-owning common carrier (NVOCC) (m) A firm that consolidates and disperses international containers that originate at or are bound for inland ports.

ocean bill of lading (ocean B/L) (m) Document indicating that the exporter will consign a shipment to an international carrier for transportation to a specified foreign market. Unlike an inland B/L, the ocean B/L also serves as a collection document. If it is a straight B/L the foreign buyer can obtain the shipment from the carrier by simply proof of identity. If a negotiable B/L is used, the buyer must first pay for the goods, post bond or meet other conditions agreeable to the seller.

on board (m) Means that cargo has been loaded on board a combined transport mode of conveyance. Used to satisfy the requirements of a letter of credit, in the absence of an express requirement to the contrary.

open account (O/A) (m) Open account, no draft drawn. Transaction payable when specified, i.e., RJM: return mail, E.O.M.—end of month; 30 days–30 days from date of invoice, 2/10/60—2 percent discount for payment in 10 days, net if paid 60 days from date of invoice. If no term is specified, O/A usually implies payment by return mail.

opening bank (m) The bank that issues the letter of credit and that makes payment according to the conditions stipulated (also known as an issuing bank).

open register (m) *See flag of convenience register.*

order bill of lading (t) *See bill of lading.*

order bill of lading (m) Usually, "to order" bills of lading are to the order of the shipper and endorsed in blank, thereby giving the holder of the B/L title to the shipment. They may also be to the order of the consignee or bank financing the transaction. Order bills of lading are negotiable (whereas straight bills of lading are not).

origin (m) Location where shipment begins its movement.

original bill of lading (OBL) (m) A document which requires proper signatures for consummating carriage of contract.

origin carrier (t) *See initial carrier.*

ORM (other regulated material) (a) *See hazardous materials and restricted articles.*

outbound (m) Cargo moving from shipper to the rail ramp. Generally refers to cargo going onto a stack or conventional train.

out gate (m) That transaction or interchange that occurs at the time a container is delivered from a rail terminal or water port to another carrier.

overage (t) An excess over the quantity believed to have been shipped or more than the quantity shown on shipping document.

overcharge (t) To charge more than the proper tariff amount.

over freight (m) Freight separated from its waybill and bearing no identifying marks. *See astray freight.*

over freight (t) Freight separated from its waybill and bearing no identifying marks. *See astray freight.*

over the bill (t) Freight in excess of that specified by the freight bill or the bill of lading.

over short and damage (OS&D) (t) Discrepancies between freight on hand and freight shown on the bill. Freight not covered by billing is "over." If some is missing it is "short." Freight received in bad condition is "damaged." Freight agents file an OS&D report showing these discrepancies.

over without bill (t) when a terminal has freight without its bill of lading or freight bill.

over-the-road (OTR) (t) Intercity.

Over-the-Road (OTR) Department (t) Department concerned with intercity operation (as contrasted to pickup and delivery service.)

over the counter small package service (a) *See small package service.*

owner (m) *See also managing owner, registered owner, parent company.* Ambiguous term, which can mean either the registered owner or the parent company which actually controls such matters as the purchase and disposal of the ship. In all cases, the owner is the individual or firm with financial equity or which is considered to be at risk if the ship is lost or libeled. In practical terms, the owner would be the beneficiary of any insurance policy on the ship, but we are rarely in a position to know who those entities are. Owner of ships need not be firms or individuals, at all, but can be countries or ministries of countries. Thus, the registered owner of a Cuban flag-of-convenience ship could be Rose Islands Steamship Company. The parent company might then be Empresa Mambisa. The true owner is the government of Cuba because it owns Mambisa, which owns Rose Islands.

packing list (m) Itemized list of commodities with marks/numbers but no cost values indicated. Usually a warehouse document.

packing improper (t) Any packing that does not comply with the classification rules and regulations for proper packing.

packing list (t) A detailed specification of packed goods.

padlock seal A "self-locking" seal that looks like a padlock. The seal locking body (head) is often made of plastic, with the shackle being made of metal. The head usually has a serial number and sometimes a logo or company name. Despite the term, padlock seals are usually not meant to function as locks. Padlock seals with plastic locking bodies are often used on commercial and residential utility meters.

pallet (a) A platform with a flat metal framed undersurface on which goods are assembled and secured by nets and straps.

parcel post air freight (a) An airline service through which a shipper can consolidate a number of parcel post packages (with destination postage affixed by the shipper) for shipment as air freight to the postmaster at another city for subsequent delivery within local postal zones or beyond.

parent company (m) Also referred to, somewhat inaccurately, as a beneficial owner. The ultimate ship-owning level in a string of firs, the lowest of which owns individual ships. Thus, properly used, a parent company owns other companies. The parent company may be owned by large conglomerates which have shipping as subsidiary interests, or by the government of countries.

participating carrier (tariff) (t) A transportation line that uses a tariff issued by another transportation line or by a tariff publishing agent.

passive seal A tamper-indicating device that uses no electrical power or batteries (internal or external) when in regular use. For some passive seals, electrical power may be temporarily required to install, inspect, or check its status.

payload (m) The revenue-producing part of the cargo.

penalty (m) An amount of money assessed against an offending party for the violation of a rule or practice.

performance bond (m) A bond issued at the request of contractors to support their commitment to perform under contract.

performance clause (m) Part of the time charter party specifying performances the ship shall be able to attain. If the ship is too slow, or burns too much fuel, the charterer is entitled to recover the cost of time lost or excess fuel consumed.

perils of the sea (m) Those causes of loss for which the carrier is not legally liable. The elemental risk of ocean transport.

perishable cargo (m) Cargo subject to decay or deterioration.

perishable freight (m) Freight subject to decay or deterioration.

permits (t) 1. Authority granted by Interstate Commerce Commission to allow motor carriers to operate in interstate commerce. 2. Permission granted to carriers by states to transport freight which exceeds legal weight and size limits.

pickup or delivery allowance (t) An allowance made by a carrier to a consignor (or owner of goods) for bringing freight to carrier's terminal. The allowance is made in lieu of carrier pickup of the freight. Delivery allowance is for consignee (or owner) pickup of goods in lieu of carrier delivery.

pickup and delivery service (PU&D) (a) An optional service for the surface transport of shipments fro shipper's door to originating air terminal and from the air terminal or destination to receiver's door. Pickup service, at an additional charge, is provided upon shipper's request. Delivery service is provided automatically by the air carrier at an additional charge unless the shipper requests otherwise. PU&D service is provided between all airports and all local points of such airports. For service beyond the terminal area see truck/air service.

piggyback (t) Transportation of a highway trailer on a rail flat car. There are five basic piggyback plans. Plan I is transportation on a trucking company owned trailer on a rail flatcar, for which the trucking company pays a negotiated rate. In Plan II, the railroad furnishes both trailer and flatcar. In Plan III, a shipper or freight forwarder owns the trailer. Plan IV is like Plan III but the shipper or forwarder also owns the flatcar. Under Plan V, the trucking company and railroad cooperate to offer a joint rate. A trucking company trailer is used.

pilferage (m) The act of stealing cargo.

piracy (m) An assault on a vessel, cargo, passengers, or crew, usually from another vessel while at sea by persons acting for personal gain and not acting on behalf of any recognized flag or international authority. Also includes acts of rioters who attach a ship from the shore for the purpose of theft or of passengers who attack the ship and its personnel from on board.

place of acceptance (m) Location where carrier actually accepts cargo from shipper or his agent.

place of delivery (m) Place where cargo leaves the care and custody of the carrier.

place of receipt (m) Location where cargo enters the care and custody of the carrier.

plastic strap seal A one-piece plastic molded strap with one end that snaps irreversibly into a head or housing on the other end, after the plastic strap is passed through the hasp of a container or door. Typically very inexpensive. Also known as ribbon seal.

POD (a) *See proof of delivery.*

port of entry (a) Any place designated by act of U.S. Congress, executive order of the President of the United States, or order of the U.S. Secretary of the Treasury, at which a U.S. Customs officer is assigned with authority to accept entries of merchandise, to collect duties, and to enforce the various provisions of the U.S. Customs laws.

port of entry (m) Port where cargo actually enters a country where the cargo is not part of its commerce.

port of entry (t) A port where foreign goods are admitted into a country. Ports of entry are officially designated by the government. Also refers to border stations maintained by some states to check truck compliance with their laws.

port of exit (m) Place where cargo actually leaves a country where the cargo is not part of its commerce.

port of loading (m) Port where cargo is transferred from one vessel to another.

point of origin (m) The station at which a shipment is received by a carrier from the shipper.

point of origin (t) The terminal at which a shipment is received by a transportation line from the shipper.

port-of-origin air cargo clearance (a) For the convenience of exporters moving goods by air from inland U.S. cities, certain U.S. Customs formalities can now be handled at the originating airport city. This avoids delaying such procedures until the export reaches a gateway point sometimes hundreds of miles from the exporter's business.

prepaid charges (a) The transportation trade practice under which the shipper pays the transportation charges.

pressure-sensitive adhesive seal *See adhesive tape seal.*

priority air freight (a) Reserved air freight or Air Express service wherein shipments have a priority after mail and the small package services. Any size or weight allowed within air freight service limits is acceptable. Advanced reservations are permitted for movement on a given flight and in some cases a partial refund is paid the shipper if the shipment is not moved on the flight specified.

priority logistics management (a) The application of the JIT (just-in-time) transportation theory.

pro forma (a) When coupled with the title of another document (proforma invoice, proforma manifest) it means an informal document presented in advance of the arrival or preparation of the required document in order to satisfy a Customs requirement.

pro forma invoice (m) An invoice provided by a supplier prior to the shipment of merchandise, informing the buyer of the kinds and quantities of goods to be sent, their value, and important specifications.

proof of delivery (POD) (a) Information provided to payor containing name of person who signed for the package with the date and time of delivery.

proof of delivery (t) Carrier establishes proof of delivery from delivery receipt copy of freight bill signed by consignee at time of delivery. This is legal proof of delivery.

proof of performance (a) *See proof of delivery.*

promotional rate (a) A rate applying to traffic under special conditions and usually confined to movement between a limited number of cities. Early rates on fresh farm produce which helped develop increased air freight volumes from the west coast to eastern cities are examples of promotional rates. *See special rates.*

proportional rate (t) A rate specifically published to be used only as a factor in making a combination through rate. A rate from New York to Chicago published to apply only to traffic destined to points beyond Chicago would be a proportional rate.

proportional tariff (t) A tariff containing only proportional rates.

protective service (a) Many airlines offer a protective service where shippers can arrange to have their shipments under carrier surveillance at each stage of transit from origin to destination. This service can be extended to pickup and delivery. Shippers can also arrange for armed guard protection. There is usually an extra charge for various levels of protective service. *See signature service.*

provisions (m) Food and beverages supplied to a ship for its crew and passengers. Included in the meaning of stores.

publication (t) Making public tariffs, circulars, billing instructions, guide books, territorial directories, classifications, exception sheets. This must be done in the manner required by the Interstate Commerce Act or state law.

radio-frequency (RF) tag A tag that emits a unique radio signal.

railbill of lading (m) A document exchanged between a shipper or shipper's agent and a railroad, which describes the required movement of a container via rail.

range of ports (m) A series of ports at which a charter party will permit a ship to discharge, one of which is to be nominated by the charterer by a specific date. Generally expressed by the names of the ports at either end of the range, e.g. the Hamburg Antwerp range. The ship will frequently be underway before the discharge port is known, and could appear to be describing an erratic track or to have given a false declaration as a result.

rate (t) The charge for transporting freight.

rate basis (t) A formula containing the specific factors used in making a rate.

rate breaking point (t) One of the places used in constructing a rate; or a place where a rate is divided.

re-consignment (m) The act of changing the bill of lading as to consignee or destination, while the shipment is in transit.

reefer container (m) Refrigerated container.

re-flag (m) To transfer the registry of an existing, registered, ship from one national authority to another. The normal procedure is to first "sell" the ships to a new registered owner under the jurisdiction of the country to which the ship is being re-flagged. That new registered owner then registers the ship under the new flag and may, or may not, depending on the requirements of the maritime laws of the countries involved, terminate the old register. At the most basic legal level, re-flagging changes the set of national laws and jurisdictions under which the ship operates. Under international law and practice the ship has the nationality of the political entity whose flag it is entitled to fly. Suits brought against the owner of the ship, for non-payment of mortgages, etc., have to be in accordance with the law of the flag state.

register (m) The record of a ship's ownership and nationality as listed with the maritime authorities of a country. Also, the compendium of such individual ships, registrations, within a country. Registration of a ship provides it with nationality and makes it subject to the laws of the country in which registered (the flag state) regardless of the nationality of the ship's ultimate owner.

registered owner (m) The owner of record of an individual ship that is listed on the registration documents (filed with a national authority).

released (t) A condition limiting the carrier's liability for loss or damage to an amount agreed upon by shipper and carrier.

released date (t) A special low rate for shipments on which the carrier's liability for loss or damage is limited to an agreed amount.

remittance following collection (RFC) (a) In instances when the shipper has performed services incident to the transportation of goods an airline will collect payment for these services from the receiver and remit such payment to the shipper. Carriers charge nominal fees for this service.

reparation (t) Compensation for damage.

reshipment (t) Goods sent to another destination under conditions which do not make the act subject to re-consignment rules and charges of the carrier.

restricted articles (a) An airline term meaning a hazardous material as defined by Title 49, Code of Federal Regulations (U.S.) and Air Transport Restricted Articles Circular 6-D. Restricted articles may be transported domestically and be classified dangerous goods when transported internationally by air. *Also see dangerous goods and hazardous material.*

restricted articles (t) commodities that cannot be handled at all or may be handled under certain specific conditions.

retroactive (t) Extending to effect matters that occurred in the past.

revocable letter of credit (m) A letter of credit that can be cancelled or altered by the drawee (buyer) after it has been issued by the drawee's bank.

sale (m) change of owner of a ship corresponding to an agreement which usually involves transfer of funds. The actual, transfer of funds could be "on paper" only, and the only change of owner required to generate a typical sale report would be that of the registered owner. Since sale does not automatically result in a change of manager or charterer, it does not automatically follow that a sale will result in a shift of control of a ship.

schedule (t) A publication filed with the Interstate Commerce Commission containing minimum charges, rules and regulations of operations of motor carriers.

sea waybill (m) A waybill is a non-negotiable document that closely resembles the B/L in form, but it is absolutely different in legal definition. Air carriers use waybills exclusively to avoid all the legal and financial obligation and complexity that are inherent in a traditional bill of lading.

seal A product designed to leave non-erasable, unambiguous evidence of entry. Unlike locks, seals may offer little or no resistance to unauthorized access. Unlike intrusion alarms, seals report entry after the fact. They must be inspected, either manually or electronically, to determine if unauthorized access has taken place. All seals have some kind of unique identifier or "fingerprint," such as a serial number or random pattern.

seal (m) Metal strip and lead fastener used for locking freight car or trailer doors. Seals are numbered for record purposes.

seal (t) A device to assure that truck doors are not opened in transit.

seal protocols The official and unofficial procedures used for seal procurement, storage, accounting, installation, inspection, removal, disposal, reporting, interpreting, and training.

seal record (m) A record of the number, condition and marks of identification on seals made at various times and places, referring to the movement of the container between origin and destination.

sectional tariff (t) A tariff made in several parts. Each section contains different rates between many of the same points, with provisions for alternative application.

secure container The tamper-indicating equivalent of a safe. The container is the seal.

security seal *See seal.*

seizure (m) In marine law includes capture or seizure by revenue customs officers of a foreign state and covers every act of taking possession of a ship. The work seizure has a stronger meaning that arrest as it may be interpreted to be a forcible possession by an overpowering force or authority of the state.

sensitive cargo Items not necessarily of intrinsic value sufficient to warrant excess valuation declaration, but of such nature that the loss, theft, tampering, or mishandling of such could entail serious public safety consequences. Sensitive cargo includes, but is not limited to: firearms and other weapons, narcotics and dangerous drugs, and secret scientific material.

sheer (m) The upward curve of a deck and lines of a ship from amidships toward bow and/or stern. A convex curve, whereas the camber will be concave.

shipment (m) The tender of one lot of cargo at one time from one shipper at one location to one consignee at one destination, on one bill of lading.

shipper (m) The person or company who is usually the supplier or owner of commodities shipped.

shipper's account number (a) *See account number.*

shipper's letter of instruction (a) A form used by a shipper to authorize an airline to issue an air waybill on the shipper's behalf. The form contains all details of shipment and authorizes the airline to sign the air waybill in the name of the shipper.

shipper's load and count (t) Indicates that the contents of a truck were loaded and counted by the shipper and not checked or verified by the transportation line.

shipper's order (t) The document authorizing release of a shipment traveling on an order bill of lading. *See bill of lading.*

shipping order (t) Instructions to carrier for transportation of a shipment. Usually it is a copy of the bill of lading. Used also as a record by the freight agent at origin.

shipping papers (t) Papers used in connection with movement of freight.

ship's manifest (m) An instrument in writing, signed by the captain of a ship, that lists the individual shipments constituting the ship's cargo.

shortage (t) When quantity actually received is less than that shown on the waybill.

sight credit (m) A letter of credit under which drafts are payable upon presentation or on demand if the documents meet the terms and conditions of the letter of credit.

sight draft (S/D) (m) A draft so drawn as to be payable on presentation to the drawee (or within a brief period thereafter, known as "days of grace"); also referred to as a demand draft. Custom, in certain areas, has in effect made an arrival draft out of a sight draft because buyers often will pay a sight draft until the arrival of the carrying vessel.

signature service (a) A service designed to provide continuous responsibility for the custody of shipments in transit, so named because a signature is required from each person handling the shipment at each stage of its transit from origin to destination.

slot charter (m) Typically used to denote hiring of space (slots) aboard a container ship by a shipper who can generate a fixed amount of cargo but one insufficient to justify taking an entire ship on hire. This type of charter will not give the charterer any control over the movement and employment of the ship. Also, less typically, used to designate a specific voyage charter to replace a temporarily unavailable ship.

small package service (a) A specialized service to guarantee the delivery of small parcels within specified express time limits, e.g. same day or next day. This traffic is subject to size and weight limitations. Air carriers that also transport passengers will accept these packages at the airport ticket counters with delivery at destination baggage claim area. Many carriers provide door-to-door service on a 24-hour basis.

space charter (m) The taking on hire of part of ship's capacity. Unlike a slot charter, the space charter does not necessarily imply a long-term relationship. This type of charter will not give the charterer any actual control over the employment and movement of the ship.

special rates (a) Rates that apply to traffic under special conditions and usually at a limited number of cities. Examples of such rates are container rates, exception ratings, surface-air rates, and import rates.

special service tariff (t) A tariff containing charges and rules governing extra services such as switching, storage, demurrage, re-consignment and diversion.

specific commodity rate (a) Rate applicable to certain classes of commodities, usually commodities moving in volume shipments.

Hence, specific commodity rates are usually lower than the general commodity rate between the same pair of cities.

stale bill of lading (m) A bill of lading which is not presented to the bank within the time frames specified in the letter of credit. Or, if no date is specified, within 21 days following the date of shipment in accordance with UCP 500.

standard international trade classification (SITC) (m) A standard numerical code used to classify commodities used in international trade.

standard rate (t) A rate established for direct routes from one point to another. Rates via other routes between the same points are set in relation to the standard rate.

standard route (t) The carrier or carriers having a direct route between two points.

standby letter of credit (m) A letter of credit issued at the request of the bank's customer ensuring payment by the bank to the beneficiary in the event the customer does not fulfill the underlying obligation and the beneficiary complies with the terms of the letter of credit.

statute of limitation (t) A law limiting the time in which claims or suits may be instituted.

statutory notice (t) The period of time required by law for giving notice of changes to be made in tariffs.

stevedore (m) Individual or firm that employs longshoremen.

stevedore (t) One who loads or unloads boats.

stiff (m) Said of a ship having a tendency to roll quickly or with a "snap" roll. Normally caused by stowage of heavy or dense material low in the ship.

stopping in transit to finish loading or unloading (m) The accessorial service of stopping volume shipments to finish loading or partial unloading at intermediate points when provided in appropriate tariffs.

stopping in transit to finish loading or unloading (t) An accessorial service of halting volume shipments to finish loading or do partial unloading at points between origin and final destination.

storage (t) Safekeeping of goods in a warehouse.

storage charges (t) Payment due for storage.

storage in transit (t) Temporary safekeeping of a shipment at a point between origin and destination.

storage vessel or station vessel (m) A merchant ship that is in use at a port to consolidate part cargoes for transfer to similar or larger ships. A storage ship can normally also get underway to carry cargo if so desired, but usually acts in a stationary capacity over a long period. Can function as a spare parts warehouse for visiting ships in addition to its primary function providing cargo storage. Provides a stationary function similar to the mobile one of lightering.

store door delivery (t) Movement of goods to a consignee's place of business.

stores (m) Supply of fresh and dry provision and the supplies for the running of the ship, such as lubricating oil, line, and spare parts.

stowage factor (m) The ratio of a cargo's cubic measurements to its weight, expressed in cubic feet/meters to the ton. Used in conjunction with the grain or bale capacities to determine how much total cargo may be loaded.

stowage, freight (t) Freight stored or packed as in a terminal or on board a truck.

stowage plan (m) A plan in the form of a longitudinal cross-section of the ship, which shows the location in the ship of all consignments of cargo, frequently color coded to highlight the various ports of discharge. The stowage plan is used by the chief mate (or first mate) who draws it up to plan discharging of the cargo. In modern container ships the stowage plan will be a computerized document, normally prepared ashore by the ships operations office.

straight bill of lading (m) A nonnegotiable bill of lading in which the goods are consigned directly to a named consignee.

straight bill of lading (t) *See bill of lading.*

stripping (m) A term often used to denote the process for removing cargo from a container.

stuffing (m) Denotes the process of loading cargo into a container.

supercargo (m) A person employed by a ship owner, charterer of a ship, or shipper of goods to supervise cargo handling operations and/or the security of the cargo in transit. During Desert Shield/Storm military Supercargo personnel often accompanied shipments of weapons and ammunition to the Kuwait Theater of Operations.

supplement (tariff) (t) Publication containing additions or charges in a tariff.

supra (t) Term denoting that a case or ruling has been previously cited in full.

SWIFT (m) Society for Worldwide Interbank Financial Telecommunications. This international system has been established to move funds and information more efficiently among member banks.

TACM (a) Transit air cargo manifest.

T&E (transportation and exportation (a) An entry filed to control the movement of merchandise being transported through the United States fro exportation.

tag A product that can be used to fingerprint, recognize, or classify an object or container at a later date, to avoid confusing it with similar looking object or container, and to detect counterfeiting. A tag may also be a unique intrinsic characteristic of the object (or container), or else a marker placed on or in the object (or container).

tamper-evident packaging The packaging contains the seal or is the seal. Often involves frangible foils, brittle films, plastic wrap, pop-up pressure bubbles, and break-off caps, lids, or tabs designed to indicate if the package has been opened. Often used on consumer products; required on over-the-counter drugs.

tamper-indicating device (TID) *See seal.*

tamper-indicating packaging *See tamper-evident packaging.*

tamper-indicating seal *See seal.*

tare weight (a) The weight of the container before loading of goods being shipped; the actual weight of the container when empty.

tariff (a) A document setting forth applicable rules, rates, and charges for the movement of goods. A tariff sets forth a contract of carriage for the shipper, the consignee, and the carrier. In addition to the domestic tariffs published by Airline Tariff Publishing Company some airlines also publish their own tariffs covering special services. International tariffs containing freight rates of the U.S. international carriers are published by the U.S. flag carriers.

tariff-motor carrier (t) The publication of a common carrier setting forth rates and charges for services it will perform for the shipping public. Copies of tariffs are filed with the Interstate Commerce Commission when the service is interstate in character and with the state regulatory body when the service is entirely intrastate.

tariff circular (ICC) (t) A circular issued by the Interstate Commerce Commission prescribing rules and regulations to be observed by transportation lines in publishing tariffs.

temporary importation under bond (TIB) (a) Temporary admission into the United States under a conditional bond for articles not imported for sale or for sale on approval. *See carnet.*

tender (m) Said of a ship with a tendency to roll slowly, or to hang on the outer edges of a roll. Normally caused by stowage of heavy or dense material high up in the ship. Although a tender ship has a more comfortable roll than one that is stiff, it is the stiff ship, which has the greatest stability, as demonstrated in its ability to right itself.

terminal (t) A building for the handling and temporary storage of freight as it is transferred between trucks (from a city pickup to an intercity truck, for example).

terminal carrier (t) The line haul motor carrier making delivery of a shipment at its destination.

theft (m) Feloniously taking and removing property with intent to deprive the rightful owner; the taking of the entire container or article. *See also pilferage.*

through bill of lading (a) A single bill of lading covering both the domestic and international carriage of an export shipment. An air waybill, for instance is essentially a through bill of lading used for air shipments. Many ocean shipments require two separate documents: an inland bill of lading for domestic carriage and an ocean bill of lading for international carriage. Through bill of lading, therefore, cannot be used.

through bill of lading (m) A single bill of lading converting both the domestic and international carriage of an export shipment.

TIB (a) *See temporary importation under bond.*

TID A tamper-indicating device.

time charter (m) Hire of a ship and its crew for a set period. Subject to any restrictions within the charter party the charterer will determine what cargo will be carried, and to which ports. The charterer is responsible for the ship's bunkers, payment of cargo handling and port charges. The owner continues to pay the crew, and hull and machinery insurance and remains responsible for the navigation and technical operation of the ship.

time definite delivery (a) The range of service performance standards offered by air freight carriers which permit the customer to select a specific time frame for delivery based on requirement for service and economy. These service standards provide door-to-door (pickup and delivery) schedule patterns based on same day, next day, second or third day delivery needs.

time draft (m) A draft maturing at a certain fixed time after presentation or acceptance. Maturity may be a given number of days after sight (acceptance) or a given number of days after date of draft.

tolerance (t) Permissible variation. For example, some states allow a tolerance in their maximum truck weight limits.

ton mile (a) One of the most commonly used measures of transportation service. In air cargo a ton mile means one ton of cargo flown one mile.

trace (m) The act of locating a shipment that has not arrived at a destination. It is also the routine check of the location of a given container on a railroad using the PC-based railroad tracing system.

tracer (m) A request upon a transportation line to trace a shipment in order to expedite its movement, establish delivery, or determine its location. 2. A request for an answer to a communication, or for advice concerning the status of a subject/particular load.

tracer (t) A request that a carrier locate a shipment to speed its movement or to establish delivery. Or a request for an answer to a previously filed claim or other communication.

trade acceptance (m) A draft drawn by the seller of goods on the buyer and accepted by the buyer. It often includes a statement indicating that the acceptor's obligation arises out of the purchase of goods from the drawer of the draft.

trailer interchange receipt (TIR) (m) A form used by the parties delivering or receiving containers and container equipment. It is used for equipment control and damage purposes. Synonymous with equipment interchange receipt (EIR)

transaction statement (m) A document that delineates the terms and conditions agreed upon between the importer and exporter.

transferable letter of credit (m) A letter of credit addressed to the beneficiary "and/or transferee (or transferees)," enabling the beneficiary to transfer the credit to another party.

transship (m) To transfer goods from one transportation line to another or to transfer goods from one vessel to another.

transshipment (m) A term used when shipment to a destination is best reached by reshipping the goods form another port.

transit air cargo manifest (TACM) (a) Procedures under which air cargo imports move through the gateway city to the city of final U.S. Customs destination for the collection of duty and other import processing, thereby expediting shipment movements, reducing gateway congestion, and saving expense for importers, the Customs Bureau, and the airlines.

transport terminal (m) A railway terminal, a freight station, a container terminal or yard, a multi-purpose cargo terminal or any similar receiving point.

transportation (m) The movement of traffic from one place to another.

Transportation Act of 1920 (t) An Act of Congress, approved February 29, 1920 providing for the termination of federal control of transportation lines at midnight, February 29, 1920, and amending the Act to Regulate commerce.

trim (m) 1. The relationship between a ship's draught forward and aft. If a ship cannot be maintained level on an even keel, it is safer to trim the ship down slightly by the stern. This serves to keep waves from breaking over the forepart of the ship, and more fully submerges the propeller and rudder, increasing their effectiveness. The term "trimmed by the bow (or head)" and "trimmed by the stern" mean that the bow or stern are loser in the water, respectively. 2. To move cargo or adjust ballast to affect the ship's trim. To level a bulk cargo in a ship's hold to maximize its stability while at seal.

truck/air service (a) the surface movement of air freight to and from airports and origin and destination points beyond the terminal area of pickup and delivery service. Directory listing cities served is available through your local airline office.

twenty-foot equivalent unit (TEU) (m) The common unit used indicating capacity of a container vessel or terminal.

ULD (unit load device) (a) Term commonly used when referring to containers and pallets.

unclaimed freight (m) (t) Freight which has not been called for by the consignee or owner, or freight that cannot be delivered.

unclean bill of lading (m) One in which a notation has been made by the carrier of any defects found in the goods when they are received for transporting. For example, such phrases as "3 hogsheads broken" or "4 sacks torn" may be inserted.

undercharge (t) To charge less than the proper amount.

uniform customs and practice for documentary credits (m) The rules and guidelines agreed to by representatives of the private sector engaged in international trade, for the conduct of trade covered by letters of credit. Generated under the sponsorship of the International Chamber of Commerce, the rules have been published as the "Uniform Customs and Practice for Documentary Credits (1993 Revision) International Chamber of Commerce Publication No. 500." The effective date is January 1994.

unitization (a) The practice or technique of consolidating many small pieces of freight into a single unit.

usance (time) credit (m) A letter of credit that calls for payment against drafts at some specified date in the future. Usance letters of credit allow buyers payment terms of a specified number of days usually not longer than six months.

valuation, actual (t) Actual value of goods required to be shown on bill of lading by shipper, when rate to be applied is dependent on that fact.

valuation charges (a) Transportation charges assessed shippers who declare a value of goods higher than the value of carriers limits of liability. *See declared value for carriage.*

valuation, released (t) Value of goods set by shipper as the limit of carrier's liability in consideration of rate to be charged.

voyage charter (m) Contract of carriage for use of a ship (or part of a ship's capacity) for one or more consecutive voyages where the ship owner remains in full control of the ship agreeing to carry a set cargo from a port, or area, to a port, area, or range of delivery ports. The ship owner pays all costs, while collecting from the charterer a fee based on tonnage carried, or a lump sum irrespective of the amount of the ship's capacity used. This form of charter leaves maximum control of the ship in the hands of its owner, as opposed to the bareboat charter, which transfers all essential control to the charterer.

vulnerability assessment Discovering and demonstrating ways to defeat a security device or security program. A thorough vulnerability assessment may include suggesting counter-measures and security improvements.

war risk (m) Insurance coverage for loss of goods resulting from any act of war.

warehouse (m) A place for the reception, delivery, consolidation, distribution, and storage of goods/cargo.

warehouse (t) A place for the reception and storage of goods.

warehouse receipt (t) A receipt for goods placed in a warehouse (may be issued as a negotiable or non-negotiable document).

warehouse receipt (m) In general, a receipt for commodities deposited with a bonafide warehouseman which identifies the commodities deposited. A warehouse receipt on which it is stated that the commodities referred to thereon will be delivered to the depositor or to any other specified person or company is a non-negotiable warehouse receipt; a warehouse receipt on which it is stated that the commodities will be delivered to the "bearer" or to the order of any specified person or company is a negotiable warehouse receipt. Endorsement (without endorsement if used to the order of "Bearer") and delivery of a negotiable warehouse receipt serves to transfer the property covered by the receipt. There is no difference between a warehouse receipt and a bill of lading in this respect.

warehouseman (t) A person in the business of receiving goods and merchandise to be stored in his warehouse.

warehousing (m) The storing of goods/cargo.

warehousing (t) The storing of goods.

Warsaw Convention (a) An international multilateral treaty which regulates, in a uniform manner, the conditions of international transportation by air. Among other things, it establishes the international liability of air carriers and establishes the monetary limits for loss, damage, and delay.

waybill (WB) (m) A document prepared by a transportation line at the point of a shipment; shows the point of the origin, destination, route, consignor, consignee, description of shipment and amount charged for the transportation service. It is forwarded with the shipment or sent by mail to the agent at the transfer point or waybill destination. Unlike a bill of lading, a waybill is *not* a document of title.

waybill (t) Description of goods sent with a common carrier freight shipment.

weather deck (m) Uppermost continuous deck of a ship; does not include forecastle, superstructures, or poop. Not necessarily the main or freeboard deck.

weight break (a) Levels at which the air freight rate per 100 pounds decreases because of substantial increases in the weight of the shipment. Examples of levels at which weight breaks occur (in pounds) are 100, 500, 1,000, 3,000, 5,000, and 10,000.

weight cargo (m) A cargo on which the transportation charge is assessed on the basis of weight.

weight certificate (m) Certifies cargo weight and measurements; issued by an official weigher on the dock, then entered on the back of the second and third copies of the dock receipt by the clerk on the dock.

weight list (m) A list showing the weights of either individual parcels, bales, or in the case of bulk commodities, and entire cargo.

weight sheets (t) Itemized list furnished by shippers to weighing bureaus itemizing articles in each consignment.

wire loop seal A seal consisting of one wire twisted around one or more other wires with a metal or plastic head or housing that crimps, traps, or holds the ends of the wires.

without recourse (m) A phrase preceding the signature of a drawer or endorser of a negotiable instrument; signifies that the instrument is passed onto subsequent holders without any liability to the endorser in the event of nonpayment or non-delivery.

without reserve (m) When a shipper's agent or representative is empowered to make definite decisions and adjustments abroad.

yard (m) A classification, storage, or switching area.

yard jockey (t) Person who operates a yard tractor or yard mule, a small tractor used to move semi-trailers around the terminal yard.

yield (a) The air transport revenue derived per unit of traffic carried in air transportation. Examples: Revenue per freight ton mile and revenue per passenger mile flown.

zip code (a) A numerical code, established by the U.S. Postal Service, used for the purpose of routing and to identify delivery zones. some airlines apply this code for freight in the same manner.

zone (a) Any one of a number of sections or districts of the United States used for the purpose of establishing proper rates for parcels, mail, and pickup and delivery.

Appendix B

LOGISTIC SECURITY ORGANIZATIONS AND SOURCES OF INFORMATION

The task at hand is to establish and maintain a cargo security program providing organizational, physical, and procedural standards and information. This publication is an effort to aid industry management in this undertaking. As corollary, the U.S. Customs has issued a booklet entitled Standards for Cargo Security (09623 09485/771). In some cases, attaining and implementing these recommended standards may entail substantial expense. They are also effective. Conversely, "bargain basement" measures provide an open invitation to theft and pilferage. The false economy created by skimping on security must be eliminated. Experience demonstrates that the loss of profit dollars far outweighs the cost of effective security.

National Cargo Security Council

The National Cargo Security Council is dedicated to the improvement of cargo security practices and procedures by shippers and transport carriers of all modes in the domestic and international commerce of the United States. Its purpose is to improve cargo transportation security by providing leadership and voluntary industry efforts; to act as a central clearinghouse for the collection and dissemination of information relating to trends, techniques, and results of efforts to prevent cargo crimes; to articulate such purposes to the public in general, and to assist and support voluntary and self-help initiatives by cities, transportation centers, and industry cargo security groups. (National Cargo Security Council, #3 Church Circle, Suite 292, Annapolis, Maryland 21401-0000; phone (410) 956-0941; fax (410) 956-0679.)

The Maritime Security Council

The Maritime Security Council works with agencies of the U.S. government and other nations that are concerned with commercial vessel security, sharing the common goals of preventing illegal drug trafficking, stowaways, theft, piracy, terrorism, and hijacking. These agencies include the U.S. Customs Service, the U.S. Coast Guard, the Immigration and Naturalization Service, the U.S. Department of State, and the U.S. Maritime Administration. As a clearinghouse for the ongoing exchange of information among its members, the Council serves as a liaison between industry and government agencies, offering current information on government activities, industry meetings, hijacking and piracy attacks, stowaways, and vessel drug seizures. The Council also maintains a library of pertinent information and training aids on maritime security. (The Maritime Security Council, 2329 Acacia Road, Port Republic, MD 20676, phone (410) 586-9200; fax (410) 586-2900.

International Association of Chiefs of Police

The International Association of Chiefs of Police is strongly committed to the cargo crime battle. Its cargo security committee has provided training and other aid to law enforcement and industry personnel involved in cargo security. The association works to develop specialized training courses through the United States Federal Law Enforcement Training Center. (International Association of Chiefs of Police, 515 North Washington Street, Alexandria, VA 22314; phone (703) 836-6767; fax (703) 836-4543.

International Association of Airport and Seaport Police

On the global scale, the International Association of Airport and Seaport Police has developed a truly international team of port law enforcement managers and directors capable of coordinating cargo theft investigations around the world. The IAASP is also committed to assisting the United Nations on the world's cargo security situation, and has offered its services, expertise, and other resources to help all nations deal with the problem. (International Association of Airport and Seaport Police, 580-2755 Lougueed Highway, Port Coquitiam, British Columbia, Canada V3B 5Y9; phone (604) 942-2132; fax (604) 942-1755.

Technology Asset Protection Association

The Technology Asset Protection Association (TAPA) is a non-profit corporation formed by the security professionals of computer and consumer electronics companies to address security issues specifically in the high-tech industry. TAPA's membership includes manufacturers and distributors of computers, computer components, and consumer electronics. TAPA's first priority is to address the challenge of losses due to theft and misappropriation of high-tech product while in transit. TAPA has embarked on several initiatives to reduce these losses. One of the initiatives is to certify transportation services providers to ensure that they have implemented the minimum-security guidelines recommended by TAPA. The Chairman of TAPA is Steve Lund of Intel Corporation. He may be contacted at (602) 554-5036. You may rite to him at 145 S. 79th Street, CH10-95, Chandler, AZ 85226.

International Chamber of Commerce

The International Chamber of Commerce, with the support of commercial interests worldwide, embarked on a major program to prevent and control marine fraud and related suspect practices. The culmination of their work was the establishment of the International Maritime Bureau in 1980. The IMB's objectives are to receive information provided by commercial and other interests relating to fraudulent or suspect practices and to collate this information as a basis on which to determine appropriate measures, to suggest avenues of procedure to those who are involved in the transaction which they suspect may be fraudulent, to contribute to provision of advice in order to improve operational and commercial systems to reduce their vulnerability to suspect practices, to contribute to the design and provision of educational services relating to the above, and to provide a reference point concerning relevant information. (International Chamber of commerce, 156 Fifth Avenue, Suite 308, New York, New York; phone (212) 206-1150; fax (212) 633-6025.

Information on the Internet

Air and Expedited Motor Carriers' Conference
www.logcity.com

Air forwarders' Association
www.logcity.com

Air Transport Association of America
www.air-transport.org

American Association of Port Authorities
www.aapa-ports.org

American Federation of Railroad Police
www.afrp

American Institute of Marine Underwriters
www.marineunderwriters.com

American Trucking Associations
www.trucking.org

American society for Industrial Security-International
www.asisonline.org

Association of Certified Fraud Examiners
www.acfe.org

Association of American Railroads
www.aar.org

Bureau of Alcohol, Tobacco & Firearms
www.atf.treas.gov

Bureau of Transportation Statistics
www.bts.gov

Canadian Association of Chiefs of Police
www.cacp.ca

Canadian Council of Motor Transport Administrators
www.ccmta.ca

Canadian International Freight Forwarders' Association
www.ciffa.com

Canadian Institute of Traffic and Transportation
www.citt.ca

Canadian Transport Web
www.cteam.ca/ctw

Chamber of Maritime Committees
www.cmc-ccm.com

Check Information Systems
www.anglefire.com/biz/BOGUS/index.html

Central Intelligence Agency
www.ic.gov

Chamber of Commerce of the United States
www.uschamber.org

Department of Defense Security Institute
www.dtic.mil/dodsi

Drug Enforcement Administration
www.usdoj.gov/dea/deahom

Federal Aviation Administration
www.faa.gov

Federal Bureau of Investigation
www.fbi.gov

Federal Railroad Administration
www.fra.dot.gov

Federal Trade Commission
www.ftc.gov

Inland Marine Underwriters Association
www.imua.org

International Association of Chiefs of Police
www.theiacp.org

International Association of Refrigerated Warehouses
www.iarw.org

International Chamber of Commerce
www.chamber.org

International Council of Shopping Centers
www.icsc.org

International Foundation of Protection Officers
www.ofpo.com

Marine Underwriters
www.marineunderwriters.com

National Defense Transportation Association
www.ndta.volpe.dot.gov

National Warehouse Association
www.whfa.org

The Private Investigators Mall
www.pimall.com

The Fraud Information Center
www.echotech.com

National Association of Police Organizations
www.napo.org

National Computer Security Association
www.ncsa.com

National Association of State Aviation Officials
www.nasao.org

NATSO (National Association of Truck Stop Operators)
www.natso.com

National Maritime Museum
www.sunymaritime.edu.museum

National Transportation Library
www.dot.gov/ntl.htm

Occupational Safety and Health Administration
www.osha.gov

Police links:
www.leolinks.com
www.officer.com
www.interpol.com

Railway Progress Institute
www.rpi.org

Seacompanion
www.Seacompanion.com

The Counter Terrorism Page
www.interlog.com/~vabrio

The International Air Cargo Association
www.tiaca.org

The Private Investigators Mall
www.pimall.com

The Transport Web
www.transportweb.com

Transport Canada Information Service
www.tc.gc.ca

Transportation Association of Canada
www.tac-atc.ca

Transportation Safety Board of Canada
http://bst-tsb.gc.ca

U.S. Customs Service
www.customs.ustreas.gov

U.S. Department of Defense
www.defenselink.mil

U.S. Department of State
www.state.gov

U.S. Department of Transportation
www.dot.gov

U.S. Maritime Administration
www.marad.dot.gov

U.S. Marshall Service
www.usdoj.gov/marshalls

U.S. Merchant Marine Academy
www.usmma.edu

U.S.-Mexico Chamber of Commerce
www.usmcoc.org

U.S. Nuclear Regulatory Commission
www.nrc.gov

U.S. Secret Service
www.treas.gov/usss

Volpe National Transportation Systems Center
www.volpe.dot.gov

Worldwide Transportation Solutions, Inc.
www.wwtsi.com/governme.htm

Appendix C

CARGO SECURITY CHECKLIST

This list of security checks may be used for many different types of facilities. It is designed to enable the facility management to select those elements pertaining to their particular security situation when devising a tailor-made survey.

- ❑ Is the perimeter of the facility defined by a fence or another physical barrier form?
- ❑ If a fence is used, does it meet the minimum specification?
    - ❑ Is the top guard strung with barbed wire and angled outward and upward at a 45 degree angle?
    - ❑ Is it at least 10 feet high?
    - ❑ Is it located so that it is not adjacent to mounds, piers, docks, cargo, handling equipment, or any other aid to surmounting it?
- ❑ If any part of the building forms a part of the perimeter or barrier, does it provide security equivalent to that provided by chain link fencing? Are all openings properly secured?
- ❑ If a masonry wall or building forms a part of the perimeter barrier, does it meet minimum specifications of perimeter fencing?
- ❑ If a river, lake, or other body of water forms any part of the perimeter barrier, are security measures equal to the deterrence of a 10-foot fence provided?
- ❑ Are openings such as culverts, tunnels, manholes, and sidewalk elevators which may permit access to the facility secure?
- ❑ List the number, location, and physical characteristics of all perimeter entrances.

- ❑ Are all portals in the perimeter barriers guarded, secured, or under constant surveillance?
- ❑ Are all perimeter entrances equipped with secure locking devices, and are they locked when not actively used?
- ❑ Are gates and/or other perimeter entrances which are not in active use frequently inspected by security officers or responsible personnel?
- ❑ Are all normally used pedestrian and vehicle gates effectively and adequately lighted so as to ensure proper identification of individuals and examination of credentials?
- ❑ Are appropriate signs setting forth the provisions for entry conspicuously posted at all principal entrances?
- ❑ Are clear zones maintained on both sides of the perimeter barrier? If the clear zone requirements cannot be met, what additional security measures have been implemented?
- ❑ Are automobiles permitted to park against or too close to perimeter barriers?
- ❑ What is the frequency of checks made by maintenance crews relative to the condition of perimeter barriers and gates.
- ❑ Do security officers regularly patrol perimeter areas?
- ❑ Are reports of inadequate perimeter security immediately acted on?
- ❑ Are bumper or fence guards to prevent damage by vehicles in place?
- ❑ If damages are reported, are necessary repairs immediately effected?
- ❑ Are perimeters protected by intrusion alarm devices?
- ❑ Does any new construction require installation of additional perimeter barriers or additional

Lighting

- ❑ Is the perimeter of the installation protected by adequate lighting?
- ❑ Are the cones of illumination from lamps directed downward and away from the facility proper and guard personnel?

- ❑ Are lights mounted to provide a strip of light both inside and outside the fence line?
- ❑ Are lights checked for proper operation periodically and any inoperative part replaced immediately?
- ❑ Do light beams overlap to provide coverage in case a bulb burns out?
- ❑ Is additional lighting provided at vulnerable or sensitive areas?
- ❑ Are gate guard posts provided with proper illumination?
- ❑ Are light finishes or strips used on lower parts of buildings and structures to aid security officer observation?
- ❑ Does the facility have a dependable auxiliary source of power?
- ❑ Is there alternate power for the lighting system independent of the plant lighting or power system?
- ❑ Is the power supply for lights adequately protected? How?
- ❑ Is the standby or emergency equipment tested periodically?
- ❑ Is emergency equipment designed to go into operation automatically?
- ❑ Is wiring tested and inspected periodically to ensure proper operation?
- ❑ Are multiple circuits used? If so, are proper switching arrangements provided?

Alarms

- ❑ Is an alarm system used in the facility?
- ❑ Is the system UL approved?
- ❑ If multiple systems are in use, are the systems integrated?
    - ❑ Does the system indicate an alert only within the facility?
    - ❑ Does the alarm signal in a central station?
    - ❑ Is it connected to facility security officer headquarters?
    - ❑ Are all openings protected?

- ❑ Is it connected directly to an enforcement headquarters outside the facility proper? Is it a private protection service? Police station? Fire station?

❑ Is there any inherent weakness in the system itself?

❑ Is the system supported by properly trained and alert security officers?

❑ Is the system for operating areas turned off during working hours?

❑ Is the system tested prior to activating it during periods of inactivity?

❑ Is the system inspected and tested regularly?

❑ Is the system tamper resistant? Weatherproof?

❑ Is an alternate system provided in the event of failure of the primary alarm?

❑ Is an alternate or independent source of power available for use?

❑ Is the alarm system properly maintained by trained personnel?

❑ Is a 24-hour repair service on call?

❑ Are periodic tests conducted to determine the adequacy of response to alarm signals?

❑ Are records kept of all alarm signals received to include time, date, location action taken, and cause for alarm?

Communications

❑ Is the security communications system adequate?

❑ What means of communication are used?

Telephone

- ❑ Is it a commercial switchboard system? Independent switchboard?
- ❑ Is it restricted for security officer use only?
- ❑ Are switchboards adequately guarded?
- ❑ Are there enough call boxes and are they conveniently located?

- ❑ Are open wires, terminal boxes, and cables frequently inspected for damage, wear, sabotage, and wiretapping?
- ❑ Are emergency phone numbers attached to all phones in the facility?

Radio

- ❑ Is proper radio procedure practiced?
- ❑ Is there an effective routine code being used?
- ❑ Is proper authentication required?
- ❑ Is the equipment properly maintained?

Teletype

- ❑ Is an operator available at all times?

Public Address

- ❑ Does it work?
- ❑ Is it audible throughout the facility?

Electronic

- ❑ there more than one system of communications available for the exclusive use of Have emergency procedures been established?
- ❑ Is the security communications equipment in use capable of transmitting instructions to all key posts simultaneously?
- ❑ Does the equipment allow a security officer to communicate with management with minimum delay?
- ❑ Is the e-mail communication secure?
- ❑ Is security personnel?
- ❑ Does one of these systems have an alternate or independent source of power?
- ❑ Has the communications center been provided with adequate physical security safeguards?

Personnel Identification and Control

- ❑ Is an identification card or badge used to identify all personnel within the confines of the facility? Controlled areas only?
- ❑ Is the identification medium designed to provide a desired degree of security?
- ❑ Does the identification in the control system include arrangements for the following:
  - ❑ Protection of the meaning of coded or printed components of badges and passes?
  - ❑ Designation of the various areas requiring special control measures to which the badge holder may be authorized entrance?
  - ❑ Strict control of identification data?
  - ❑ Clear explanation and description of the identification data used?
  - ❑ A clear statement of the authorization and limitation placed on the bearer?
  - ❑ Details of where, when, and how badges should be worn?
  - ❑ Procedures to be followed in case of loss or damage to identification media?
  - ❑ Procedures for recovery and invalidation?
- ❑ If a badge exchange system is enforced for any restricted area, does the system provide for:
  - ❑ Comparison of badge or pass with the personnel?
  - ❑ Physical exchange of restricted area badge or general authorization badge at time of entrance and exit?
  - ❑ Recording each badge exchange?
  - ❑ Inventory of badges issued by security personnel at the start and completion of tours of duty?
  - ❑ Location of personnel who have not checked out of the area at the close of each tour of duty?
  - ❑ Security of badges not in use?

- ❑ Are messengers who are required to traverse areas of varying degrees of security provided with special identification?
- ❑ Are the prescribed standards for access to exclusion areas supplemented with arrangements as follows:
  - ❑ At least one representative of management or security is in the area at all times when work is in progress?
  - ❑ No other persons are permitted to enter the area until one representing management or security has entered?
  - ❑ The management or security representative remains until all others have departed the area?
- ❑ Are personnel who require infrequent access to a critical area and who have not been issued regular security identification treated as visitors and either issued a visitor's pass or badge or a special one-time pass?
- ❑ Are all personnel required to wear the security identification badge while on duty?
- ❑ Do security officers at control points compare badges of bearers both upon entry and exit?
- ❑ Is supervision of personnel charged with checking identification badges sufficient to ensure continuing effectiveness of the identification and control system?
- ❑ Are badges recorded and controlled by rigid accountability procedures?
- ❑ Are lost badges replaced with one bearing a different number or one that is not otherwise identical to the one lost?
- ❑ Are procedures relative to lost, damaged, and/or forgotten badges adequate?
- ❑ Are temporary badges issued?
- ❑ Are lists of lost badges posted at all security officer control points?
- ❑ Are badges of such design and appearance as to readily enable security officers and other personnel to recognize the authorizations and limitations applicable to the bearers both quickly and accurately?

- ❑ How long ago were currently used badges originally issued?
- ❑ Do existing procedures ensure the return of identification badges upon termination of employment? Are badges similar to or identical to employee badges issued to outside contractor employees or special visitors? E.g., vendors, trades persons, or utility or equipment service persons?
- ❑ Have local regulations governing identification and control been revised in any material way since first established?
- ❑ Are all phases of the system under supervision and control of a security office?
- ❑ Is an effective visitor escort procedure established?
- ❑ Are visitors required to conspicuously display identification at all times while in the facility?
- ❑ When visitors leave the facility, are they required to turn in their identifications badges, and is the departure time in each case recorded on a visitors' register or log?
- ❑ What procedures are invoked when visitor identification badges are not turned in prior to departure?
- ❑ What measures are employed, other than issuance of identification badges, to control the movements of personnel from transportation companies working within the perimeter of the facility?
- ❑ Is enforcement the most vulnerable part of an identification system?

Package and Material Control

- ❑ Are there standard procedures relative to control of packages and materials?
- ❑ Are all security officers conversant with these control measures?
- ❑ Are notices on restriction and control procedures prominently displayed in each active entrance and exit?
- ❑ Is there a check room or separate area where employees and visitors can leave their packages?
    - ❑ Is an adequate receipt system in effect?
    - ❑ Are packages inspected in the owner's presence?

- ❑ Is access to the checking area restricted to authorized personnel only?
- ❑ Is a policy established for disposition of items left beyond a specified period?

❑ Are spot checks of persons and vehicles conducted, and if so, are frequency and scope established?

❑ Are detection devices used?

❑ Is a property removal slip, signed by assigned personnel, required when property is being removed from a facility?

❑ Are removal slips available in the security office for signature by these personnel authorizing property removal?

❑ Are property removal slips surrendered to security officers at exit points?

❑ Are special rules established for package and material handling?

- ❑ Are package and material passes used to exempt bearers from search?
- ❑ Are time, date, bearer's name, agency, and description of contents properly recorded?
- ❑ Is preparation and issuance rigidly controlled?
- ❑ Are the passes serially numbered?
- ❑ Do the rules provide for signature of validating personnel?
- ❑ Are signature cards readily available to security officers for comparison?

❑ Is an effective procedure in use for control and search of special vehicles?

- ❑ Emergency vehicles?
- ❑ VIP vehicles?
- ❑ Special courier vehicles?
- ❑ Vendors vehicles?

Vehicle Control

- ❑ Are vehicles which are allowed regular access to the facility registered with the security office?
- ❑ Have definite procedures been established for the registration of private vehicles, and are they issued in writing?
- ❑ Do the vehicle registration requirements apply also to motor vehicles owned or operated by employees of any individual, firm, corporation, or contractor whose business activities require use of vehicles on the facility?
- ❑ Are annual or more frequent registrations required?
- ❑ What information is incorporated into the registration application forms?
- ❑ Do the prescribed prerequisites for registration include a valid vehicle registration and operator's license?
- ❑ Is mechanical inspection of vehicles and/or proof of financial responsibility required as a prerequisite of authority to operate a vehicle within the facility?
- ❑ Are decals or metal permit tags affixed to all vehicles authorized to operate within the facility?
- ❑ Do registration permits bear a permanently affixed serial number and numerical designation of year of registration?
- ❑ Do the regulatory controls for registration include:
    - ❑ Prohibition against transfer of registration permit tags for use with a vehicle other than the one for which originally issued?
    - ❑ Replacement of lost permit tags at the registrant's expense?
    - ❑ Return of tags to the security office when the vehicle is no longer authorized entry into the facility?
    - ❑ Destruction of invalid decals or metal tags?
- ❑ What is the nature and scope of registration records maintained by the security office?
- ❑ Do the gate security officers make periodic checks to ensure that vehicles are operated on the premises only by properly licensed persons?

- ❑ Is the specified system used to control for the movement of commercial vehicles into and out of the facility?
- ❑ Are loading and unloading platforms, located outside the operating areas, separated one from the other and controlled by security officer supervised entrances?
- ❑ Are all trucks and other conveyances required to enter through gates manned by security officers?
- ❑ If trucks are permitted direct access to operating areas, are the driver's cab and vehicle's content carefully examined?
- ❑ Does the check at entrances cover both incoming and outgoing vehicles?
- ❑ Are escorts provided when vehicles are permitted access to operating or controlled areas?
- ❑ Does the supervision of loading operations insure that unauthorized cargo does not enter or leave the facility via trucks or other conveyances?
- ❑ Are temporary tags issued to visitors' vehicles?
- ❑ Are automobiles allowed to be parked within operating or controlled areas?
- ❑ Are parking lots provided? Adequate? Lighted?
- ❑ Are interior parking areas located away from sensitive points such as cargo handling and document processing areas?
- ❑ Are interior parking areas fenced so that occupants of automobiles must pas through a pedestrian gate when entering or leaving the working area?
- ❑ Are separate parking areas provided for visitors' vehicles?
- ❑ Are there restrictions against employees entering private vehicle parking areas during duty hours?

Lock and Key Control

- ❑ Are the locks and keys to all buildings and entrances controlled by a designated key control person?

- ❑ Does the key control person have overall responsibility for issuance and replacement of locks and keys?
- ❑ Are keys issued only to authorized personnel?
- ❑ Is the removal of keys from the premises prohibited?
- ❑ Are keys not in use secured in a locked, fireproof cabinet?
- ❑ Are current records maintained indicating:
  - ❑ Clear record of person to whom key is issued?
  - ❑ Time of issue and return of keys?
  - ❑ Buildings and/or entrances to which keys are issued?
  - ❑ Number and identification of key or keys issued?
  - ❑ Location and number of master keys?
  - ❑ Location and number of duplicate keys?
  - ❑ Location of keys and locks held in reserve?
- ❑ Is a current key control directive in effect and understood?
- ❑ Are locks changed immediately upon loss or theft of keys?
- ❑ Are inventories and inspections conducted by the key control designee to ensure compliance with directives? How often?
- ❑ If master keys are used, are they devoid of markings identifying them as such?
- ❑ Are losses or thefts of keys promptly investigated by the key control person?
- ❑ Must all requests for reproduction or duplication of keys be approved by the key control designee?
- ❑ Are locks on inactive gates in storage facilities under seal? Are they checked periodically by guard personnel?
- ❑ Are locks rotated within the installation at least semiannually?
- ❑ Where applicable, is the manufacturer's serial number on combination locks obliterated?
- ❑ Are measures in effect to prevent the unauthorized removal of locks on open cabinets, gates, or buildings?

Security Officers

- ❑ Is a security officer force provided? Is it responsive to management?
- ❑ Indicate authorized and actual strength broken down by positions.
- ❑ Have there been changes since the last survey to the authorized or actual security officer forces?
- ❑ Is the present security officer force strength commensurate with the degree of security protection required?
- ❑ Is the use of security officer forces reviewed periodically to ensure effective and economical use?
- ❑ Is supervisor responsibility for security officer force operations vested in a security office?
- ❑ Is a security officer headquarters provided?
- ❑ Does the security officer headquarters contain controlled equipment and instruments of all alarm, warning, and guard communication systems? Are the equipment, instruments, and communication systems adequate?
- ❑ Are security officers familiar with communication equipment used?
- ❑ Does security headquarters have direct communication with local municipal fire and police departments?
- ❑ Do security officers meet the minimum qualification standards?
- ❑ Are they certified?
- ❑ Is each security officer required to complete a course of basic training and take periodic courses of in-service or advanced training to cover;
    - ❑ Common forms of theft, pilferage, and sabotage activity?
    - ❑ Types of bombs and explosives?
    - ❑ Location of hazardous materials and processes?
    - ❑ Location and use of fire protection equipment including sprinkler control valves?

- ❑ Location and operation of all important steam and gas valves and main electrical switches?
- ❑ Conditions which may cause fire and explosions?
- ❑ Location and use of first air equipment?
- ❑ Duties in the event of fire, explosion, natural disaster, civil disturbance, power blackout, or HAZMAT release?
- ❑ Use of the communications system?
- ❑ Proper methods of search?
- ❑ Methods of observation and description?
- ❑ Methods of patrolling terminals, warehouses, and other areas within the facility?
- ❑ Supervision of visitors?
- ❑ Report writing, log and record keeping?
- ❑ General and special guard orders?
- ❑ Cargo handling operations?
- ❑ Documentation processing?

❑ Are security officers equipped with uniforms that are complete, distinctive, and authoritative in appearance?

❑ Are periodic examinations conducted to ensure maintenance of security officer training standards?

❑ Are activities of the security officers in consonance with established policy? Is there a formal manual covering standard operating procedures?

❑ Is supervision of the security officers adequate?

❑ Are general and special orders properly posted?

❑ Are security officer orders reviewed at least semiannually to assure applicability?

❑ Are periodic inspections and examinations conducted to determine the degree of understanding and compliance with all security officer orders?

- ❑ Do physical, functional, or other changes at the installation indicate a necessity for, or feasibility of establishing additional guard posts or discontinuing any existing posts or control?
- ❑ Is two-way radio equipment installed on all security officer patrol cars?
- ❑ Are duties other than those related to security performed by security officer personnel?
- ❑ Do security officers record or report their presence at key points in the facility by means of portable watch clock, central watch clock stations, telephones, or two-way radio equipment?
- ❑ Are security officer assignments and patrol routes varied at frequent intervals to obviate an established routine?

Documentation

- ❑ Are all security officers familiar with specific guidelines and procedures established for cargo operations and documentation?
- ❑ Are all shipping documents controlled and properly stored?
- ❑ Is there an accountable audit trail to follow cargo movements?
- ❑ Are positive checks made of all cargo received from shippers or members of the transportation network prior to signing any receipts?
- ❑ Are the documents provided with proper date, time, and signature notations?
- ❑ Is there a cargo loss reporting system in effect that will provide specific information on losses?
- ❑ Does this system contribute to the development of recommendations to prevent cargo loss?
- ❑ Are detailed procedures for identifying and handling over, short, and damaged cargo (OS&D) developed for training and reference by all personnel involved in the handling and movement of cargo?
- ❑ Is anyone designated as the OS&D responsible person?
- ❑ Are formal OS&D reports prepared?

- ❑ Are these reports complete and the information provided specific, e.g., notations for shortage reflecting number of items found rather than the number assumed to be missing?
- ❑ Are the loss recording procedures implemented immediately upon discovery of an overage, shortage, or damage to a shipment?
- ❑ What are the procedures for handling short shipments? Is the on-hand cargo moved to destination awaiting recovery of the remainder? If so, is a free astray waybill prepared for the balance?
- ❑ Is a central clearinghouse or office maintained for all losses within the system?
- ❑ Are periodic reports prepared that include information on a value of loss commodities, trend, geographical areas where losses are occurring, specific cargo difficulties, ongoing shipper errors, and any other poor procedures that have been identified? If so, are proposed corrective actins included in these reports?
- ❑ On containerized shipments, are seal numbers recorded on the cargo documentation?
- ❑ Are tracing procedures established for high valued/target items?

Special Cargo

- ❑ Are protected measures for the movement and storage of high valued cargo detailed in standard operating procedures?
- ❑ Is high valued cargo separated and secured?
- ❑ Is a security cage or crib utilized?
- ❑ Are high valued shipments on hand inventoried on a regular basis, preferably at the shift change?
- ❑ Is a lot maintained that lists transfers in and out of the security area to include shipment identification, waybill number, number of pieces, date in, date out, and signature service?
- ❑ Is personnel access to the high value area controlled?
- ❑ If high valued cargo is placed in containers or trailers, is it located in a special security holding area where observation by management and/or security personnel is facilitated?

- ❑ When containers are mounted on frames, is the fifth wheel secured by a king pin lock which meets the minimum standards for locks and is constructed to withstand normal abuse from equipment? Is a designated individual responsible for storage and control of these locks?
- ❑ In transit, are vehicles containing high value loads escorted? Is routing planned with security in mind?
- ❑ Are drivers trained to follow proper security measures to include hijacking avoidance measures?
- ❑ Are vehicles equipped with alarms? Are the alarms maintained?
- ❑ Are two-way radios installed in the vehicles?
- ❑ Are padlocks applied to vehicle cargo doors?
- ❑ When unattended, are parked vehicles locked?
- ❑ Are all containerized shipments sealed?
- ❑ Are all sealed containerized shipments entering or leaving the facility inspected? If the seals are not intact or there is evidence of tampering, or the seal numbers are incorrect, are the cognizant security and/or management personnel notified and a physical tally of the cargo conducted?
- ❑ Are all container movements logged in and out, showing date, time, seal number, name of truck operator and company, and the registration number of the equipment used?
- ❑ Are unused seals stored in a secured area? Is a seal distribution log maintained?
- ❑ Are seals applied and/or broken only by authorized personnel?
- ❑ Are the number of letter codings on all seals clearly visible? Are the seals used corrosion resistant and tamper proof?

NOTES

NOTES

NOTES

NOTES

NOTES

NOTES